R and MATLAB®

Chapman & Hall/CRC
The R Series

Series Editors

John M. Chambers
Department of Statistics
Stanford University
Stanford, California, USA

Torsten Hothorn
Division of Biostatistics
University of Zurich
Switzerland

Duncan Temple Lang
Department of Statistics
University of California, Davis
Davis, California, USA

Hadley Wickham
RStudio
Boston, Massachusetts, USA

Aims and Scope

This book series reflects the recent rapid growth in the development and application of R, the programming language and software environment for statistical computing and graphics. R is now widely used in academic research, education, and industry. It is constantly growing, with new versions of the core software released regularly and more than 6,000 packages available. It is difficult for the documentation to keep pace with the expansion of the software, and this vital book series provides a forum for the publication of books covering many aspects of the development and application of R.

The scope of the series is wide, covering three main threads:
- Applications of R to specific disciplines such as biology, epidemiology, genetics, engineering, finance, and the social sciences.
- Using R for the study of topics of statistical methodology, such as linear and mixed modeling, time series, Bayesian methods, and missing data.
- The development of R, including programming, building packages, and graphics.

The books will appeal to programmers and developers of R software, as well as applied statisticians and data analysts in many fields. The books will feature detailed worked examples and R code fully integrated into the text, ensuring their usefulness to researchers, practitioners and students.

Published Titles

Stated Preference Methods Using R, *Hideo Aizaki, Tomoaki Nakatani, and Kazuo Sato*

Using R for Numerical Analysis in Science and Engineering, *Victor A. Bloomfield*

Event History Analysis with R, *Göran Broström*

Computational Actuarial Science with R, *Arthur Charpentier*

Statistical Computing in C++ and R, *Randall L. Eubank and Ana Kupresanin*

Reproducible Research with R and RStudio, Second Edition, *Christopher Gandrud*

R and MATLAB® *David E. Hiebeler*

Nonparametric Statistical Methods Using R, *John Kloke and Joseph McKean*

Displaying Time Series, Spatial, and Space-Time Data with R, *Oscar Perpiñán Lamigueiro*

Programming Graphical User Interfaces with R, *Michael F. Lawrence and John Verzani*

Analyzing Sensory Data with R, *Sébastien Lê and Theirry Worch*

Parallel Computing for Data Science: With Examples in R, C++ and CUDA, *Norman Matloff*

Analyzing Baseball Data with R, *Max Marchi and Jim Albert*

Growth Curve Analysis and Visualization Using R, *Daniel Mirman*

R Graphics, Second Edition, *Paul Murrell*

Data Science in R: A Case Studies Approach to Computational Reasoning and Problem Solving, *Deborah Nolan and Duncan Temple Lang*

Multiple Factor Analysis by Example Using R, *Jérôme Pagès*

Customer and Business Analytics: Applied Data Mining for Business Decision Making Using R, *Daniel S. Putler and Robert E. Krider*

Implementing Reproducible Research, *Victoria Stodden, Friedrich Leisch, and Roger D. Peng*

Graphical Data Analysis with R, *Antony Unwin*

Using R for Introductory Statistics, Second Edition, *John Verzani*

Advanced R, *Hadley Wickham*

Dynamic Documents with R and knitr, Second Edition, *Yihui Xie*

R and MATLAB®

David E. Hiebeler

University of Maine
Orono, USA

CRC Press
Taylor & Francis Group
Boca Raton London New York

CRC Press is an imprint of the
Taylor & Francis Group, an **informa** business

A CHAPMAN & HALL BOOK

CRC Press
Taylor & Francis Group
6000 Broken Sound Parkway NW, Suite 300
Boca Raton, FL 33487-2742

© 2015 by Taylor & Francis Group, LLC
CRC Press is an imprint of Taylor & Francis Group, an Informa business

No claim to original U.S. Government works

Printed on acid-free paper
Version Date: 20150406

International Standard Book Number-13: 978-1-4665-6838-9 (Hardback)

Library of Congress Cataloging-in-Publication Data

Hiebeler, David E.
   R and MATLAB / David E. Hiebeler.
      pages cm. -- (Chapman & Hall/CRC the R series ; 30)
   "A CRC title."
   Includes bibliographical references and index.
   ISBN 978-1-4665-6838-9 (alk. paper)
   1. Multivariate analysis--Data processing. 2. R (Computer program language) 3. MATLAB. I. Title.

QA278.H54 2016
510.285'53--dc23   2015006954

Visit the Taylor & Francis Web site at
http://www.taylorandfrancis.com

and the CRC Press Web site at
http://www.crcpress.com

To my parents,
for encouraging me as I got started
with that first Apple][+.

Contents

List of Figures

List of Tables

Preface

Who this book is for

I have interacted with many people over the past several years after making my "MAT-LAB/R Reference" available on my Web site. Based on those conversations, there is a large population of people out there who have used MATLAB®[1] for some period of time, but who now find themselves working with biologists, statisticians, or some other professionals who speak R rather than MATLAB. There is a complementary group of people who use R, who now find themselves trying to work and share code with colleagues who use MATLAB. The intended reader is someone who already knows one package, and for whatever reasons, now needs or wants to learn the other.

This book grew out of the MATLAB/R Reference document mentioned above. That document is a concise reference summary of many of the key topics presented in this book, and continues to be available on my Web site, at `http://www.math.umaine.edu/~hiebeler`.

My own experience learning R

I fall into the first category of users described above. I used MATLAB for many years, primarily for prototyping my research simulations (which I would then rewrite in C for faster performance) and for data visualization and graphics. I also regularly taught a course on "Modeling and Simulation" in which I used MATLAB as the software platform. The course covers various biological models, including stochastic spatial models. I found more and more biologists signing up for the class over the years, and some of them started asking if they could use R because they had already started learning to use it in their statistics classes. I was somewhat tired of learning new programming languages/environments, and was not particularly eager to learn another one.[2] But I reluctantly decided to look into R. At first I was very frustrated with some of the differences that struck me right away — for example, when you edit a file defining a new R function, you cannot just type its name to call the function, but instead must "**source**" it first. Typing a matrix into R is certainly more tedious than doing so in MATLAB.

However, over time, I came to appreciate the power and flexibility of many features of R. The fact that it is available for free on Windows, Mac OS-X, and Linux certainly has

---

[1]MATLAB® and Simulink® are registered trademarks of The MathWorks, Inc. For product information, please contact: The MathWorks, Inc., 3 Apple Hill Drive, Natick, MA 01760-2098 USA; Tel: 508-647-7000; Fax: 508-647-7001; E-mail: info@mathworks.com; Web: www.mathworks.com.

[2]The main programming languages I had used over the years at that point were Applesoft BASIC, Apple][+ machine language, Pascal, Modula-2, Forth, C, Lisp, APL, SNOBOL, PostScript, Java, Perl, MATLAB, and a bit of Python. Plus various libraries/environments and programming-like things such as SunView, X11, csh scripting, LATEX, and HTML. That list is certainly smaller than that of many computer scientists, but long enough for me.

made it easier for the many students who work on research with me or take my courses to obtain a copy for their own computers.

There are still some times when I miss the more concise/simple way of expressing things in MATLAB (in particular, how much more convenient it is to type in a matrix), but I have now come to enjoy the time I spend working in R.

Formatting conventions and terminology

R and MATLAB commands, i.e., things you could actually type at the command prompt, are generally formatted using a `non-bold typewriter font`, for example `x = sqrt(7)`. A **bold font** is used when referring to the name of a function, file, variable, or keyword, for example the **sqrt** function, the variable **foo**, or **for** loops.

R and MATLAB are referred to in the book as "platforms," rather than the perhaps more natural "software packages," because the word "package" is a loaded word with specific meaning in R.

In some parts of the book, R and MATLAB code are placed side by side. This is done for brief commands and concepts which do not need much explanation. In other parts, R material is presented, followed by MATLAB material, with some differences between the platforms emphasized.

Commands vs. GUI

There are two main approaches to working with software platforms like MATLAB and R: doing things primarily with commands, and doing things primarily via menus through a graphical user interface (GUI). I will admit that I am old-school, and started using computers before they *had* graphical user-interfaces. Personally, I can type commands much more quickly than I can click my way to equivalent commands via menus, so I prefer the command-line approach over the GUI approach.

As one justification for my preference, using commands to achieve goals is usually a bit more portable and easier to share with others than using menus. The placements of various items in menus is sometimes different on different operating systems such as Mac OS-X vs. Microsoft Windows, so if you are trying to explain to someone how to accomplish something, and they have a different operating system than you (or even simply a different version of the software), you may have some trouble. Admittedly, some commands may also differ between operating systems or between versions of the software. But for the most part, if you write an R or MATLAB script on one operating system, it will work identically on others, and can be usefully shared with other people more easily than a description of which menu items to select and which buttons to push. The configuration and details of user-interfaces also tend to change more frequently than the commands in packages like MATLAB and R; if you tell someone how to do something using commands, it is likely to work on a wider variety of versions of the software. I therefore focus more on using commands than on clicking on menu items in this book, for those tasks where both approaches can be used.

What this book is not

This book is not intended to be a comprehensive introduction or overview of either the R or MATLAB platforms. I should warn you up front that this book will not teach you what many may consider the "best" way to do things in R or MATLAB, according to the deeper philosophy of either platform. My goal, based on my experience, is to try and show how to do things in either platform which is most similar to the ways they are done in the other, to make the transition from one platform to the other as quick and painless as possible for you. If you want to delve into the deeper ways of thinking behind using either platform, this reference can help get you started, but you should follow up with some of the other plentiful resources available.

This book is also not intended to steer you toward one platform or the other. Quite often, the choice of software platform to use depends on your context, i.e., your employer and colleagues. If you truly have a wide-open choice, it is difficult to say which platform is best for you. My personal feeling is that many things are *easier* to do in MATLAB, but can be done more *flexibly* in R, at the expense of being more complicated. That is just a general impression, and there are of course many counterexamples. Another thing to note is that The MathWorks, Inc. (developer of MATLAB) has been quite aggressive recently about improving the performance of various types of MATLAB code; many of my MATLAB computations run many times faster in newer versions of MATLAB than they used to. If you are at a large company or academic institution, you likely have access to a site license to use MATLAB. If you are looking for routines to perform specialized statistical tests, R has a vast and quickly growing library of packages to fit that need. R is of course available to download free of charge, although MATLAB can be evaluated in a free trial, and there is very affordable pricing for students or for personal use.

Acknowledgments

Thanks to my editor, John Kimmel, for being far too patient with me as I worked on this book.

Bill Halteman helped me learn R, first by going through my MATLAB simulation lab exercises and determining it would not be too difficult to implement them all in R. He then patiently answered my many questions about R (and pretty much everything else under the sun) for the following several years. Many people have also sent me suggestions and corrections for my MATLAB/R Reference that this book was originally based on. Those contributors include Juan David Ospina Arango, Berry Boessenkool, Robert Bryce, Thomas Clerc, Alan Cobo-Lewis, Richard Cotton, Stephen Eglen, Andreas Handel, Niels Richard Hansen, Luke Hartigan, Roger Jeurissen, David Khabie-Zeitoune, Seungyeon Kim, Michael Kiparsky, Isaac Michaud, Andy Moody, Ben Morin, Lee Pang, Manas A. Pathak, Rachel Rier, Rune Schjellerup Philosof, Rachel Rier, William Simpson, David Winsemius, Corey Yanofsky, and Jian Ye.

Finally, like most authors, I must thank my family and apologize to them for all of the times I chose to spend with drafts of this book rather than with them.

About the Author

David E. Hiebeler is an Associate Professor in the Department of Mathematics & Statistics at the University of Maine. He was previously a Visiting Lecturer at Cornell University. He has degrees from Rensselaer Polytechnic Institute, Harvard University, and a Ph.D. in applied mathematics from Cornell University.

He previously worked at the Center for Nonlinear Studies, the Theoretical Division, and the Advanced Computing Lab at Los Alamos National Laboratory, the Santa Fe Institute, and Thinking Machines Corporation. He began programming at age 12 on an Apple][+ computer. His research involves mathematical and computational stochastic spatial models in population ecology and epidemiology. He also dabbles in iOS (iPhone/iPad) development, primarily for K–12 outreach.

1

Installing and Running *R* and *MATLAB*

1.1 Obtaining and installing

R

R can be downloaded for free from The R Project Web site at **http://www.r-project.org**. Follow the link to CRAN (Comprehensive R Archive Network) to download R. You must first choose a CRAN mirror, i.e., one of the many Web sites around the world that keep copies of the R software. It is probably best to choose one that is geographically close to you.

Once you have chosen a mirror, you can choose which operating system (Linux, Mac OS-X, or Windows) you use. Depending on your operating system, you may then have further choices to make, such as which distribution of Linux or which version of OS-X you use. Note that the source code for R is also freely available for download from the same site.

When you actually run R on Windows or Mac OS-X, you will see a graphical user interface like the one shown in Figure 1.1. The interface primarily consists of the R Console or Command window, where you can interactively type commands for R to interpret. The standard command prompt in R is ">," which means R is waiting for you to type a command. You can try it out with some simple computations; for example, type 7*8 (and then press **Enter**) to do a simple multiplication, or `sqrt(exp(3))` to compute $\sqrt{e^3}$. When you run R on Linux in a terminal, you just get the command prompt, without the various menus in the interface for OS-X and Windows.

You may wish to use a much more comprehensive, friendlier interface, especially if you are coming to R from MATLAB. A very popular integrated development environment for R, which has many of the same useful feature's as MATLAB's interface, is called *RStudio*; it is available from `http://www.rstudio.com`. RStudio's interface is shown in Figure 1.2. If you are trying to ease your transition from MATLAB to R, you may seriously want to consider installing and loading the **pracma** package, as it implements many functions which behave nearly equivalently with MATLAB routines; see Section 13.9 for information about installing and loading packages.

MATLAB

MATLAB is available from The MathWorks, Inc. (**http://www.mathworks.com**). Pricing and licensing information is available under their "Products & Services" link. There is a very inexpensive option for students, and also special pricing for home and educational use. It is also possible to obtain a trial version of the software. When you purchase MATLAB, you will receive information for how to download and activate your copy.

MATLAB is available for all three popular operating systems (Linux, Mac OS-X, and Windows).

When you run MATLAB, you will see a graphical user interface like the one shown in Figure 1.3. The large region in the middle of the user interface is the Command Window,

```
R version 3.1.2 (2014-10-31) -- "Pumpkin Helmet"
Copyright (C) 2014 The R Foundation for Statistical Computing
Platform: x86_64-apple-darwin13.4.0 (64-bit)

R is free software and comes with ABSOLUTELY NO WARRANTY.
You are welcome to redistribute it under certain conditions.
Type 'license()' or 'licence()' for distribution details.

  Natural language support but running in an English locale

R is a collaborative project with many contributors.
Type 'contributors()' for more information and
'citation()' on how to cite R or R packages in publications.

Type 'demo()' for some demos, 'help()' for on-line help, or
'help.start()' for an HTML browser interface to help.
Type 'q()' to quit R.

[R.app GUI 1.65 (6833) x86_64-apple-darwin13.4.0]

>
```

FIGURE 1.1
The R graphical user interface in Mac OS-X.

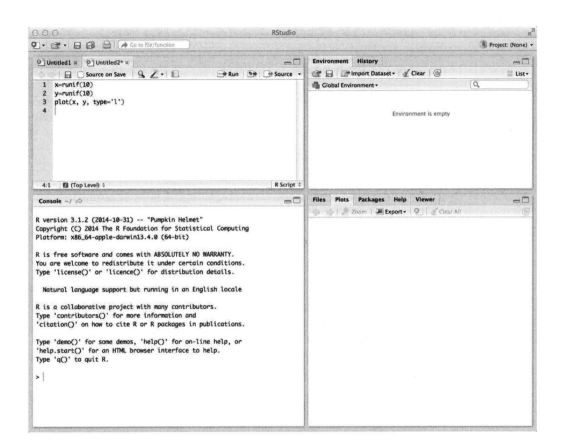

FIGURE 1.2
The RStudio graphical user interface in Mac OS-X.

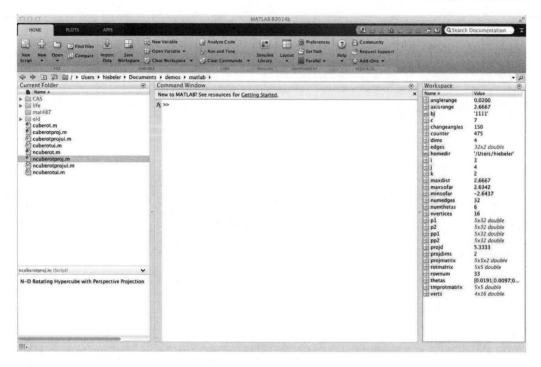

FIGURE 1.3
The MATLAB graphical user interface in Mac OS-X.

where you can interactively type commands for MATLAB to interpret. The standard command prompt in MATLAB is ">>," which means MATLAB is waiting for you to type a command. You can try it out with some simple computations; for example, type `7*8` (and then press **Enter**) to do a simple multiplication, or `sqrt(exp(3))` to compute $\sqrt{e^3}$.

There are also many other things visible in the user interface. These include:

- A file browser, listing the files in a given directory.

- More information about a selected file.

- The MATLAB workspace viewer, showing the variables that are currently defined.

1.2   Commands for getting help

Both platforms offer extensive built-in documentation. Depending on your operating system, documentation may be available from menus or other items in the user interface. Example commands are shown below, demonstrating how to access the various built-in documentation systems from the command window.

R

`help('eigen')` Display documentation for the **eigen** function.

`?eigen` Shorthand for the above command.

`?'if'` When using the shorthand form of the **help** command, if you are requesting help for a built-in keyword or punctuation, you need to surround the item with quotes. For example, you can also do `?'+'` to display help about the addition operator.

`help.search('eigenvalues')` Search for documentation matching the specified string when you cannot remember the name of a function. Enter `?help.search` for more details about options for the search.

`??eigenvalues` Shorthand for the above command. You should also enclose reserved words or punctuation in quotes, as well as multi-word phrases, e.g., `??'linear models'`.

`help('expm', package=Matrix)` or `?Matrix::expm` Search for help about **expm** specifically within the **Matrix** package. This is useful if, for example, you have loaded both the **Matrix** and **pracma** package, both of which define functions named **expm**. See Section 13.9 for more about R packages.

`help.start()` Open a browser window showing general documentation installed locally.

`RSiteSearch('binomial')` Search on-line help in a browser via the site **http://search.r-project.org**. See the "How to search" link there for more information. For example, to search for an exact phrase, use a command like `RSiteSearch('{logistic regression}')`.

`help(package='Matrix')` See the general help for the **Matrix** package.

`library(help='pracma')` Provide information about a given package (**pracma** in this example), including a list of the functions it provides.

`example(sort)` Many of the documentation pages include examples at the end; this command runs the examples from the **sort** documentation.

`data()` List the various data sets that are included with R.

`data(mtcars)` Load the **mtcars** data set containing Motor Trend road tests. You can then enter the command `mtcars` to see the raw data.

`vignette()` Display the various vignettes that are available in R. Vignettes are short pieces of documentation that are less formal than the regular "help" information. Packages may include vignettes, so which vignettes are available to you depends on which packages you have installed.

`apropos('^pr..$')` Display all four-letter objects (variable names, functions, etc.) that start with **pr**. The argument you provide can be a *regular expression* (see the help for **regex** for details about regular expressions).

MATLAB

`help('eig')` or `help eig` Display documentation for the **eig** command. When using the second form of the command, no quotes are needed around reserved keywords or punctuation; that is, `help +` and `help if` both work.

`lookfor eigenvalues` Search for documentation matching the specified string when you cannot remember the name of a function. See `help lookfor` for more details on just what this searches.

`help stats` Display help for the Statistics Toolbox.

`doc` Open the help browser.

`doc eig` Open the help browser and display information about the **eig** function.

1.3 Demos

Both platforms include mechanisms for demonstrating some of the available features, including mathematical computations and graphics. In R, enter the command `demo()` to bring up a list of available demos. You may then run specific demos from the list via commands like `demo(graphics)`, `demo(image)`, and `demo(colors)`. In MATLAB, enter the command `demo` to bring up the MATLAB browser where you can navigate a page of various demos. The page contains an extensive set of demos in a wide variety of topics such as Mathematics, Graphics, Programming, and so on.

1.4 Quitting

To exit R, you can enter the command `quit()`, or its shorter alias `q()`. R will ask whether or not you want to save your workspace (all of your data/variables), or if you wish to cancel the quit.

To exit MATLAB, enter either of the two equivalent commands `quit` or `exit`.

See Section 13.8 for more information about what happens when starting up or exiting both platforms, and how you can provide additional code to customize the startup and shutdown sequences.

1.5 Additional resources

Resources such as discussion forums about both R and MATLAB are plentiful on the Internet. You can find questions and answers at typical technology Q&A Web sites such as `http://stackoverflow.com` (just prefix your search at the site with either the string [**r**] or [**matlab**] to restrict your search to the given platform). There are also on-line resources dedicated to each platform.

R

The R Project Web site (**http://www.r-project.org**) has links to many additional resources. Under the R Project section of the site, there are links to various mailing lists and conferences about R. Under the Documentation section of the site, there are links to additional manuals, The R Journal, which has in-depth articles on various topics, a Wiki, and so on. If you navigate to a CRAN repository site (i.e., a site where you would actually download R), there is also a link to a page listing many pieces of contributed documentation. Another site to try is **http://www.rseek.org**, which performs Web searches for information specifically about R.

MATLAB

The official support page from The MathWorks (`http://www.mathworks.com/support`) has links to several helpful resources, including official documentation, a question & answer forum a file-exchange repository with many user-contributed files. There is also a link to something called Cody, a coding game which asks you to write MATLAB programs to solve specific tasks, and to do so in the most efficient way possible. The Cody problems generally have interesting discussions among the many users who have tried them. There is also a MATLAB Digest, and The MathWorks News&Notes. There are very many excellent introductions and references for MATLAB available, aimed at audiences in many different fields (engineering, digital signal processing, and so on). See References [2, 9, 11, 12, 22] for just a few examples, and see `http://www.mathworks.com/support/books/` for a list of more than 1,500 books related to MATLAB, organized by category.

2

Getting Started: Variables and Basic Computations

Performing simple computations is virtually identical in both packages. Commands like `y = sqrt(7)` and `x = sin(y)` will work in both R and MATLAB, although what you see on your screen after doing them differs between the two platforms. One difference is that in MATLAB, parentheses are optional when calling a function which needs no arguments. That is, you can enter `date` (which returns the current date) in MATLAB instead of `date()` if you like. In R, you must enter `date()` to call the analagous function.

2.1 Variable names

Variable names in R can consist of letters, digits, periods, and underscores. R installations in some locales may allow for additional characters, such as accented letters. However, for greatest portability, it is probably best to stick with simple alphanumeric characters. Variable names must start with either a letter or period; if a variable name starts with a period, the second character cannot be a numerical digit. The name "..." should also be avoided, as it has a special meaning, as will be seen in Section 8.1.5. Variable names were limited to 256 bytes before R version 2.13.0, and since then are apparently "effectively unlimited" [28]. The function **make.names** can be used to turn a questionable string into a valid variable name; for example, `make.names('7 hi there')` returns the string "**X7.hi.there**." This method also has the advantage of avoiding reserved words (it appends a period if you give it a reserved word), although it allows "..." through untouched, along with the names of built-in functions such as **sqrt**.

Variable names in MATLAB can consist of letters, digits, and underscores. The first character must be a letter. Note that the name of a variable must be less than the built-in variable **namelengthmax**, which is 63 on my reference system. Also, do *not* use periods in variable names, as they have a special meaning, namely to refer to fields of structures (see Section 3.8.3).

As seen above, variable names in MATLAB are more restrictive than they are in R, in both length and the characters allowed in them. If you will be moving between the two platforms, it may therefore be prudent to follow MATLAB's restrictions, as such variable names will be valid in both platforms. While variable names such as **N** and **X** are lacking in their descriptive abilities, I find it difficult to imagine a scenario where more than 63 characters are useful or needed for a variable name.

2.2   Assignment statements

One of the first things you will likely notice when comparing R and MATLAB code is that R has a different syntax for assigning values to variables. Although the "=" symbol can be used in both platforms (so that `y = sqrt(7)` will work, as mentioned above) and is the standard way to perform assignments in MATLAB, R has an another way to perform variable assignments. In R, one can use the syntax `y <- sqrt(7)` (or even `sqrt(7) -> y`, though this is much less common). Just to warn you, R purists eschew writing `y = sqrt(7)` in favor of the `<-` notation, and you will likely be derided at some point if you choose to use `=`. However, I still strongly prefer to simply use `=`, which is the notation used by most of the programming languages I have used in my life. It also makes it slightly less work to translate code between R and MATLAB. On the other hand, if you will be interacting with someone who uses S-PLUS, you may prefer to use `<-` to increase compatibility with them, as that is the syntax S-PLUS uses for variable assignment. I will use `=` for the assignment operator throughout this book. For the most part, `=` and `<-` may be used interchangeably in R. They are always equivalent for simple assignment statements, but they have an important difference when used in parameters to function calls (see Page 85).

Another difference between the two platforms is that R allows statements like `a=b=7` and `x = (y=7)+3` (which assign the value 7 to **a**, **b**, and **y**, and the value 10 to **x**). MATLAB does not allow you to chain together assignments in this way.

The final obvious difference between doing an assignment statement like `y = 7` is in the output displayed. Entering the command in R does not display the value. You must enter the command **y** to see the results of the computation:

```
──────────────────────── R ────────────────────────
> y = 7
> y
[1] 7
```

Notice that you do not just see the value **7**, but also get that odd-looking **[1]** at the beginning of the line. That is because technically, the value computed was a vector (of length 1), and the line of output containing the 7 begins at position 1 of that vector. If you produce a bigger vector, say with a command like `runif(13)` to produce thirteen random numbers, you will see that each line is prefixed with the position of the first value on that line:

```
──────────────────────── R ────────────────────────
> runif(13)
 [1] 0.045741256 0.957446628 0.233276158 0.645035735 0.518672549
 [6] 0.002810847 0.842379136 0.925471541 0.178368850 0.264167924
[11] 0.300473767 0.913935277 0.775792785
```

The second line of output begins with the sixth element of the vector, while the third line of output begins with the eleventh element. Personally, I found the numbers in square brackets on all of R's output very distracting at first, particularly because they are there even on strings that you output in a standard way. But I have eventually gotten used to them and even find them useful; also, there is a way to suppress them to display things in a prettier way (see Page 166). There is also another way to see the results of a computation that is normally silent; namely, you can enclose the expression in parentheses:

```
──────────────────────── R ────────────────────────
> (y = 7)
[1] 7
```

Commands in MATLAB display their results by default:

```
━━━━━━━━━━━━━━━━━━━━━━━━━ MATLAB ━━━━━━━━━━━━━━━━━━━━━━━━━
>> y = sqrt(7)
y =
    2.6458
```

which shows both the value computed, and the fact that it was stored in the variable **y**.[1] You can force MATLAB to suppress its output by appending a semicolon to the command. You can then explicitly type **y** (as in R) to see its value:

```
━━━━━━━━━━━━━━━━━━━━━━━━━ MATLAB ━━━━━━━━━━━━━━━━━━━━━━━━━
>> y = sqrt(7);
>> y
y =
    2.6458
```

See the next section for more about the difference in visible output between R and MATLAB.

2.3 Basic computations

Basic computations on scalar values can be performed directly at the command prompt in both packages. You can evaluate expressions directly, such as `sin(3)+sqrt(17)`, although you will probably more often store the results of such expressions in variables. If you evaluate an expression without storing the results into a variable (as in the example above), both platforms will show the result of the computation.[2] As explained in the previous section, if you store a value into a variable, R does not display the value, while MATLAB does.

You can put multiple commands on a single line. In R, simply separate the commands with semicolons, such as `x=7; y=8`. You can end a line containing a single R command with a semicolon too; the semicolon will have no effect. This is useful when translating MATLAB code to R; if you accidentally forget to delete some trailing semicolons, they are harmless. In MATLAB, it is a bit more complicated. You can suppress the output from whichever commands you like on the line, by using semicolons and commas as separators. A command followed by a semicolon has its output suppressed, while a command followed by a comma does not. Observe the following commands and their output from MATLAB:

```
━━━━━━━━━━━━━━━━━━━━━━━━━ MATLAB ━━━━━━━━━━━━━━━━━━━━━━━━━
>> x=7; y=8;
>> x=7, y=8;
x =
    7
>> x=7; y=8
y =
    8
>> x=7, y=8
```

---

[1] Note that before entering the displayed commands, I had previously entered the MATLAB command `format compact` to get rid of the extra white space around MATLAB's output.

[2] There are exceptions to this in R. When executing commands in certain ways, such as within loops or from a file, R is less verbose. See Section 7.3 for how to get around this problem. Some functions also return their values "invisibly"; see Section 8.1.1.

```
x =

        7
y =

        8
```

A comma at the end of a line (as in `x=7, y=8,`) has no effect and is unnecessary, while a semicolon at the end of a line suppresses the output from the final command.

If you want to break up an expression across multiple lines of input, the two platforms handle this differently. R will automatically detect that an expression is not complete at the end of one line, and change its ">" prompt to "+" to let you know that it is waiting for more input to complete the expression:

```
——————————————————————— R ———————————————————————
> x = 5 *
+ 8
> x
[1] 40
```

(Note that the "+" before the 8 above is the prompt provided by R; it is not user input.)

MATLAB will not automatically detect that an expression is incomplete and prompt you for more; instead, it gives an error. However, if you end a line with three dots ("..."), then it lets you continue an expression on the following line:

```
——————————————————————— MATLAB ———————————————————————
>> x = 5 *
 x = 5 *
        |
Error: Expression or statement is incomplete or incorrect.

>> x = 5 * ...
8
x =

       40
```

The advantage of MATLAB's approach is that you can break up an expression in places where R would not know that you want to continue typing:

```
——————————————————————— MATLAB ———————————————————————
>> x = 5 * 8 ...
/ 3
x =

    13.3333
```

The advantage of R's approach is that the behavior is more automatic, without the need for an explicit sign from you that input will be continued. However, if you want to split the expression `x = 5 * 8 / 3` up, you need to do it in such a way that R knows more code is coming for the expression; otherwise, it thinks you are done. If you really want to break up an expression in a place that would confuse R, you can always put parentheses around the entire thing, to let it know when you are truly done:

```
——————————————————————— R ———————————————————————
> x = 5 * 8
> / 3
Error: unexpected '/' in "/"
> x = 5 * 8 /
```

Description	R/MATLAB command
$a + b, a - b, ab, a/b$	`a+b, a-b, a*b, a/b`
$\sqrt{a}$	`sqrt(a)`
a^b	`a^b`
absolute value $|a|$	`abs(a)`
e^a	`exp(a)`
$\ln(a)$	`log(a)`
$\log_2(a), \log_{10}(a)$	`log2(a), log10(a)`
$\sin(a), \cos(a), \tan(a)$	`sin(a), cos(a), tan(a)`
$\sin^{-1}(a), \cos^{-1}(a), \tan^{-1}(a)$	`asin(a), acos(a), atan(a)`
Two-argument arctangent	`atan2(y,x)`
$\sinh(a), \cosh(a), \tanh(a)$	`sinh(a), cosh(a), tanh(a)`
$\sinh^{-1}(a), \cosh^{-1}(a), \tanh^{-1}(a)$	`asinh(a), acosh(a), atanh(a)`
Sign of a (-1, 0, or 1, according to whether a is negative, zero, or positive)	`sign(a)`
Round a to nearest integer	`round(a)`[3]
Largest integer not greater than a (round down)	`floor(a)`

TABLE 2.1
Basic computations which are identical in R and MATLAB.

```
+ 3
> x
[1] 13.33333
> x = (5 * 8
+ / 3)
> x
[1] 13.33333
```

Both platforms will automatically store the results of a recent computation in an internal variable, but with a subtle difference. R stores the results of the last expression in the variable **.Last.value**. For example, if you enter the commands 4+8; x=15 then **.Last.value** will contain the value 15. In MATLAB, the results of the last expression which were not explicitly stored in a variable will be stored in the variable **ans**. So after entering the commands 4+8; x=15, **ans** will contain 12 rather than 15, because the value 15 was stored in **x**.

Most of the simple functions and operators typically used with real numbers have the same names in R and MATLAB, with those names being common to many if not most programming languages. Table 2.1 shows such commands which are identical in the two platforms.

Both platforms include special values to represent infinity; for example, this comes up when you try to compute 1/0. In R, you can enter the value infinity as Inf; in MATLAB, you can type either **inf** or **Inf**, though MATLAB itself will always display the latter. Both platforms also include the special value NaN, which represents "Not a Number." This arises

---

[3]R uses the IEC 60559 standard, rounding 5 to the even digit. So for example, in R, `round(3.5)` and `round(4.5)` both return 4. In MATLAB, those two commands return 4 and 5, respectively.

Description	R command	MATLAB command
Smallest integer not less than a (round up)	ceiling(a)	ceil(a)
Round a toward zero	trunc(a)	fix(a)
n MOD k (modulo arithmetic, the remainder when dividing n by k)	n %% k	mod(n,k)

TABLE 2.2
Basic computations which are different in R and MATLAB.

when computing, for example, 0/0 or Inf*0. In R, you can enter this value as NaN; in MATLAB, you enter either nan or NaN.

Both platforms also provide ways to check whether values are finite, infinite, or NaN. In R, you can use is.finite(x), is.infinite(x), and is.nan(x); in MATLAB, you can use isfinite(x), isinf(x), and isnan(x). In both platforms, these functions also work with vectors and arrays/matrices.

R also has another value, called NA. This is a special value representing missing data, or the absence of values. Many statistical routines and other functions will either ignore NA or treat them specially. You can type NA to enter the value NA. The expression is.na(x) can be used to check whether a given value is NA. However, note that is.na(x) will also return TRUE if x is **NaN**.

2.4   Formatting of output

Although internal computations are done using the full precision of whatever data types are involved (such as double-precision floating point values), you can control how values are displayed in both platforms.

In R, the command options(digits=6) requests that values be displayed using 6 digits. The default is 7 digits; you can use getOptions('digits') to see the current setting.

In MATLAB, the **format** command can be used to control how values are displayed. Some of its possible usages are as follows:

- format short: scaled fixed point format with 5 digits

- format long: 15 digits for double and 7 digits for single

- format shorte: floating point with 5 digits

- format longe: floating point with 15 digits for double and 7 digits for single

- format shortg: MATLAB chooses the best of fixed or floating point format with 5 digits, according to the magnitude of the value

- format longg: best of fixed or floating point format, with 15 digits for double and 7 digits for single

The **format** command can take other arguments as well, such as for banking format or hexadecimal. Also, as mentioned in the footnote on page 11, the command format compact tells MATLAB not to pad its output with so much vertical white space. The command format loose enables the extra white space again.

2.5 Other computations

Several other miscellaneous functions are listed below, along with the commands to calculate them in both platforms.

1. The error function $\text{erf}(x) = (2/\sqrt{\pi}) \int_0^x e^{-t^2} dt$

R	MATLAB
`2*pnorm(x*sqrt(2))-1`	`erf(x)`

2. The complementary error function $\text{cerf}(x) = (2/\sqrt{\pi}) \int_x^\infty e^{-t^2} dt = 1\text{-erf}(x)$

`2*pnorm(x*sqrt(2),lower=FALSE)`	`erfc(x)`

3. The inverse error function

`qnorm((1+x)/2)/sqrt(2)`	`erfinv(x)`

4. The inverse complementary error function

`qnorm(x/2,lower=FALSE)/sqrt(2)`	`erfcinv(x)`

5. Binomial coefficient $\begin{pmatrix} n \\ k \end{pmatrix} = n!/(n!(n-k)!)$

`choose(n,k)`	`nchoosek(n,k)`

6. Factorial, $n!$

`factorial(n)`	`factorial(n)`

7. Gamma function $\Gamma(x) = \int_0^{\inf} t^{x-1} e^{-t} dt$

`gamma(x)`	`gamma(x)`

Note that R's **gamma** function returns **NaN** when x is either zero or a negative integer, while MATLAB's **gamma** function returns **Inf**.

8. Log gamma function $\ln(\Gamma(x))$

`lgamma(x)`	`gammaln(x)`

9. Beta function $\beta(z,w) = \int_0^1 t^{z-1}(1-t)^{w-1} dt = \Gamma(z)\Gamma(w)/\Gamma(z+w)$

`beta(z,w)`	`beta(z,w)`

10. Log beta function $\ln(\beta(z,w))$

Description	R command	MATLAB command
Modulus (magnitude)	`abs(z)` or `Mod(z)`	`abs(z)`
Argument (angle)	`Arg(z)`	`angle(z)`
Complex conjugate	`Conj(z)`	`conj(z)`
Real part of z	`Re(z)`	`real(z)`
Imaginary part of z	`Im(z)`	`imag(z)`
Construct complex number $a + bi$ from variables **a** and **b**	`complex(real=a,` `imaginary=b)` or `complex(1,a,b)`	`complex(a,b)`

TABLE 2.3
Basic functions for working with complex numbers.

`lbeta(z,w)`	`betaln(z,w)`

2.6   Complex numbers

Both R and MATLAB can deal with complex numbers. To assign the value $i = \sqrt{-1}$ to the variable **x**, you cannot simply enter `x=i`, as the variable **i** may already have been given another value. However, you can enter `x=1i` in both platforms. Similarly, the value $3+4i$ can be entered as `3+4i`. If you have real variables **a** and **b** and want to construct the complex number $a + bi$, you cannot simply write `a+bi`, because **bi** will be interpreted as a variable name. You could instead write `a+1i*b`, or see Table 2.3 for another way.

Table 2.3 displays some functions specifically for working with complex numbers. Most of the functions and operators from Tables 2.1 and 2.2 will also work with complex numbers, with the following exceptions. The modulo operator does not work in either platform. `sign(z)` does not work in R, while in MATLAB it is equivalent to `z/abs(z)`, i.e., it returns a value with unit modulus/magnitude and the same argument/angle as **z**. Finally, `atan2(y,z)` with complex arguments in MATLAB will not work.

2.7   Strange variable names in R

Although guidelines for variable names in R were given in Section 2.1, those rules really are just a convenience for R's command parser. It turns out you can make variable names consisting of any valid string, regardless of how strange that string is. For example, you can make a variable named "**17 is nice!**". You can store the value 18 in this variable via the command `assign('17 is nice!', 18)`. You can then access the contents of the variable via `get('17 is nice!')`. You can build up more complex expressions in this way, e.g., `assign('odd name', get('odd too') + get('enough already!'))`. I'm not sure why you might want to create such strange variable names, but the option is there.

2.8 Data types

2.8.1 R

The standard data types in R or *modes* of data, are numeric, complex numbers, logical (i.e., boolean TRUE/FALSE), character, and raw (which allows you to work with raw bytes). You can find out the mode of an object such as a variable **x** via the command `mode(x)`. Objects also have a *class*, which can be looked up via `class(x)`. For simple scalars and vectors, the class is the same as the mode. However, some objects have a different mode, such as **matrix**, **array**, **data.frame**, **factor**, and so on. Matrices, arrays, and data frames will be covered in Chapter 3. In addition to the above information, the command `typeof(x)` displays the internal storage mode of **x**.

Factors

A *factor* in R is a data type for representing ordered or unordered (nominal) categorical values. They are somewhat like an **enum** variable in the C programming language, in that they let you equate a set of scalar values (such as 1, 2, 3, etc.) with a more meaningful set of labels, so that you can use the latter labels when analyzing data.

The function **factor** can convert a vector into this special data type. For example:

```R
> v=c('red','red','blue','red','blue')
> fv = factor(v)
> fv
[1] red   red   blue red   blue
Levels: blue red
> c(fv)
[1] 2 2 1 2 1
> summary(fv)
blue  red
   2    3
> str(fv)
 Factor w/ 2 levels "blue","red": 2 2 1 2 1
```

The vector **v** contains five values, with two unique values, called *levels*. We convert this into a factor stored in the variable **fv**. When you look at **fv**, you see the five values and the two levels displayed. Internally, **fv** is equivalent to the vector containing the five values 2, 1, 1, 2, and 1. The **summary** function indicates how many elements of **fv** there are for each level, while **str** shows the internal structure of **fv**.

Note that **factor** sorts the levels alphabetically by default. If for some reason you want them in a different order, you can provide a vector specifying the labels in the desired order:

```R
> fv = factor(v,levels=c('red','blue'))
> fv
[1] red   red   blue red   blue
Levels: red blue
> c(fv)
[1] 1 1 2 1 2
> summary(fv)
 red blue
   3    2
```

```
> str(fv)
 Factor w/ 2 levels "red","blue": 1 1 2 1 2
```

To specify that a factor represents ordered categorical data, use the **ordered=TRUE** argument to factor:

─────────────────────────────── R ───────────────────────────────
```
> w=c('M','XL','M','XL','L','L')
> fw=factor(w,ordered=TRUE,levels=c('M','L','XL'))
> fw
[1] M  XL M  XL L  L
Levels: M < L < XL
> c(fw)
[1] 1 3 1 3 2 2
> summary(fw)
 M  L XL
 2  2  2
> str(fw)
 Ord.factor w/ 3 levels "M"<"L"<"XL": 1 3 1 3 2 2
```

Certain statistical routines operate differently on unordered versus ordered factors.

2.8.2   MATLAB

In MATLAB, the basic data types are logical (boolean TRUE/FALSE), double and single (for floating-point values), signed integer values of various lengths (int8, int16, int32, and int64, where the numeric suffix denotes how many bits per value are used for storage), unsigned integer values of various lengths (uint8, uint16, uint32, and uint64), and character. The standard data structure is a matrix (indeed, scalar values are simply 1×1 matrices). Other data structures include cell arrays (see Chapter 5) and structures (see Section 3.8.3). You can also store inline functions or function handles in MATLAB variables.

3

Matrices and Vectors

3.1 Overview

Both MATLAB and R can work directly with vectors and with $m \times n$ matrices (with the usual convention that m specifies the number of rows and n the number of columns). Both platforms support vectorized computations, i.e., expressions that implicitly operate on entire vectors or matrices. However, there are some important differences between the two.

- A key difference in syntax is that R uses square brackets to delimit matrix subscripts (e.g., A[2,3]), while MATLAB uses parentheses (e.g., A(2,3)). R's syntax has the advantage here because when reading MATLAB code, it is not immediately apparent whether g(2,3) is accessing an element of a matrix g, or calling a function g with two parameters.

- In R, a *vector* is a primitive data type, and a *matrix* (which is actually a special case of an *array*) is another distinct type. A vector with n elements is neither a row vector nor a column vector, but simply an ordered set of n values. Care must sometimes be taken to ensure that a value is stored as a vector or matrix in particular.

  In MATLAB, the fundamental data type is a *matrix*. A scalar value is simply a 1×1 matrix, while a vector with n elements is either an $n \times 1$ matrix (column vector) or a $1 \times n$ matrix (row vector). Care must sometimes be taken to ensure that a vector is a row vector or a column vector in particular.

- In R, you can use array subscripting on expressions. For example, to access the element in row 1, column 2 of the matrix A^2, you can do (A%*%A)[1,2]. You cannot perform the corresponding command (A*A)(1,2) in MATLAB. The easiest way around this is to use a temporary variable, e.g., tmp=A*A; tmp(1,2), although there is another way (see Section 3.9).

3.2 Creating vectors

In R, a vector is an ordered sequence of values; it is neither a row vector nor a column vector. In MATLAB, a vector is simply a special case of a matrix which has either one row or one column.

Below are many of the common ways of creating vectors.

1. Create a vector containing the specified values 4, 8, 15, 16, 23, 42

R	MATLAB
`c(4,8,15,16,23,42)`	Either `[4,8,15,16,23,42]` or `[4 8 15`
Or you can call `scan()`, then enter the	`16 23 42]` for a row vector. For a column
values separated by spaces, and press	vector, `[4;8;15;16;23;42]` or
Return/Enter twice to signal the end of	`[4`
the values. To explicitly construct a row	`8`
vector (stored as a matrix), use	`15`
`cbind(4,8,15,16,23,42)`; for a column	`16`
vector, use `rbind(4,8,15,16,23,42)`.	`23`
	`42]`

(R): The often-used function **c** "combines" its arguments; it can work with many data types.

(MATLAB): When entering matrices, commas or spaces can be used as delimiters between columns. Semicolons or newlines can be used between rows.

2. Vector of length k containing all zeros

`rep(0,k)` or `numeric(k)`	`zeros(k,1)` (for column vector) or `zeros(1,k)` (for row vector)

3. Vector of length k containing the value j in all positions

`rep(j,k)`	`j*ones(k,1)` or `repmat(j,k,1)` (for column vector); `j*ones(1,k)` or `repmat(j,1,k)` (for row vector)

(R): **rep** is a widely used function which replicates its argument. It can work with various data types, and can be used in a few different ways with vectors.

(MATLAB): **repmat** is used to replicate a matrix; because a scalar is simply a 1×1 matrix, it can also be used as above.

4. Sequence with unit increment: 7, 8, 9, 10, 11

`7:11`	`7:11` (for row vector), or `(7:11)'` (for column vector).

(MATLAB): The apostrophe transposes a matrix (but see entries 66 and 67), and can also be used to produce column vectors instead of row vectors for the entries below.

5. Values $j - 1$ through $k + 1$ with unit increment

`(j-1):(k+1)`	`j-1:k+1`

The fact that the colon has higher order of precedence than arithmetic operators in R but not in MATLAB is a major nuisance. I have seen many people struggle to find why their code is broken, only for it to be caused by this issue. In R, writing `j-1:k+1` really means `j-(1:k)+1` which is almost never what was intended. To be extra-cautious, I suggest always using parentheses when writing expressions like this in either platform.

6. Sequence with increment -1: 11, 10, 9, 8, 7

`11:7`	`11:-1:7`

If you have gotten used to doing things like `11:7` in R, be careful when using MATLAB, because it produces an empty matrix, and does not display any warnings or explicit signs of trouble.

7. Sequence with specified increment: 4, 7, 10, 13

`seq(4,13,by=3)` or simply `seq(4,13,3)`	`4:3:13`

8. n equally spaced values from a to b

`seq(a,b,length.out=n)` or this can also be shortened to `seq(a,b,len=n)`	`linspace(a,b,n)`

For entries 7 and 8, note that in R, the same function (**seq**, which generates a regular sequence of values) is used whether you are specifying the increment between values or the total number of values desired. In MATLAB, two different methods are used for these two cases.

9. The vector $1, 2, \ldots, n$, where n is the length of vector **v**

`seq_len(length(v))`	`1:length(v)`

Just writing `1:length(v)` in R will fail if **v** is an empty vector, i.e., a vector of length zero. That is because `1:0` in R produces a vector of length 2, with the values 1 and 0. Using `seq_len` avoids this problem, and produces an empty vector in R if **v** is empty. In MATLAB, because `1:0` produces an empty matrix, the above command works whether **v** is empty or not.

10. n logarithmically equally spaced values from 10^a to 10^b

`10^seq(a,b,len=n)`	`logspace(a,b,n)`

11. The values 1 through n, but excluding any values which are in the vector **v**

`(1:n)[is.na(match(1:n,v))]`	`z=1:n; z(~ismember(1:n,v))`

See entry 28 for more info about how/why these commands work.

12. Build a vector by making 3 copies of the vector **v** end-to-end

`rep(v,3)`	`repmat(v,1,3)` if **v** is a row vector; `repmat(v,3,1)` if **v** is a column vector. Or, `v(repmat(1:length(v),1,3))` works with both row and column vectors

13. Build a vector by repeating each element of vector **v** 3 times

`rep(v,each=3)`	`inds=repmat(1:length(v),3,1);` `v(inds(:))` This works with both row and column vectors

14. Build a vector **w** of length **n** by making copies of the vector **v** end-to-end

`w=numeric(n); w[1:n]=v` Note: a warning will be displayed if **n** is not an integer multiple of the length of **v**, but the commands will still work.	`k=length(v); m=floor(n/k);` `r=mod(n,k); inds=[repmat(1:v1,1,m)` `1:r]; w=v(inds)`

This is what is known as "recycling" of values in R; see entry 24 for more information on how this works.

15. Build a vector **w** of length **n** by repeating each element of vector **v** 3 times

`w=numeric(n);` `w[1:n]=v[rep(1:length(v),each=3)]`	First use the commands in entry 13 to produce a vector **u** which has each element of **v** repeated 3 times. Then use the commands in entry 14 to replicate copies of **u** to build a vector of length **n**.

In both platforms, if repeating each element of **v** 3 times produces a vector of length less than **n**, then the produced values "start over again" to produce a vector long enough.

3.3   Working with vectors

16. Access or set the third element of vector **v**

R	MATLAB
`v[3]` or `v[3]=17`	`v(3)` or `v(3)=17`

This is one of the most immediately obvious differences between R and MATLAB, that R uses square brackets and MATLAB uses parentheses when accessing matrix or vector elements. R's syntax is better here, since in MATLAB, `v(3)` could represent either the third element of vector **v**, or a call to a function **v**. When translating code from MATLAB to R, it is easy to miss places where you have forgotten to make this change. If you have a function and a vector with the same name, the error may be difficult to track down.

See item 37 for what happens if you store a value past the end of a vector.

Also, be aware that in MATLAB, `v(0)` will give an error. In R, `v[0]` gives "nothing," or more precisely, a vector of length zero (with no error or warning).

17. Length of vector **v**

`length(v)`	`length(v)` Because this is equivalent to `max(size(v))`, it works with both row and column vectors.

18. Ensure that **v** is a column vector, whether it was previously a row or column vector

`cbind(v)` turns **v** into a matrix with one column	`v=v(:)` or `v=reshape(v,length(v),1)`

19. Ensure that **v** is a row vector, whether it was previously a row or column vector

`rbind(v)` turns **v** into a matrix with one row	`v=v(:)'` or `v=reshape(v,1,length(v))`

20. Last entry of **v**

`v[length(v)]`	`v(end)`

21. Combine vectors **v** and **w** by concatenating them

`c(v,w)`	`[v w]` if they are row vectors, or `[v ; w]` if they are column vectors

22. Reverse the order of elements in vector **v**

`rev(v)`	`v(end:-1:1)`

(MATLAB): This produces either row or column vector results, corresponding to whether the original was a row or column vector.

23. A vector consisting of the fourth, first, and second elements of **v**

`v[c(4,1,2)]`	`v([4 1 2])`

(MATLAB): This produces either row or column vector results, corresponding to whether the original was a row or column vector.

24. Replace the second, fifth, sixth, and tenth element of **v** with specified values

`v[c(2,5,6,10)] = c(17,42,77,68)`	`v([2 5 6 10]) = [17 42 77 68]`

A major feature in R is what is called "recycling," where if you do not provide enough values, the values you do provide will be repeatedly re-used from the beginning as many times as necessary. So if you do something like `v[3:8] = c(50,51)` to try to set six values in **v** but only provide two values, it would be as if the right-hand side were **c(50,51,50,51,50,51)**. If the number of values needed is not an integer multiple of the number of values provided, a warning is displayed, but the command still executes (some of the values will be used more than others in that case). This recycling behavior is very common in R.

MATLAB does not have this recycling feature, and will simply give an error if the incorrect number of values is provided. The exception is that a single (scalar) value will be recycled as many times as needed; the scalar is said to be "promoted to a vector" in that case.

25. Elements of **v** from position a through the end

`v[a:length(v)]`	`v(a:end)`

26. Given a vector **b** containing logical values, create a vector giving the indices of the TRUE elements

`which(b)`	`find(b)`

Note that R's **which** function only works with logical values, not numerical values. MATLAB's **find** will work with numerical values as well, treating non-zero values as TRUE. The R equivalent of MATLAB's `find([6 0 7])` would be `which(c(6,0,7) != 0)`. The above constructs are often used with logical vectors built from expressions involving a vector, as in the next entry.

27. The indices of the positive elements of vector **v**

`which(v > 0)`	`find(v > 0)`

When testing elements of a vector, the various logical comparison operators that can be used are shown in Table 3.1. Note that they are all identical except for the "NOT" operator, which is ! in R and ~ in MATLAB.

28. Given a vector **b** containing logical (TRUE/FALSE) values, build a new vector by extracting only those elements of vector **v** whose corresponding elements of **b** are TRUE

`v[b]`	`v(b)`

This idea is often used to extract only elements from a vector **v** satisfying a certain condition, as in the next entry.

(R): If **b** is shorter than **v**, its elements will be recycled. If **b** is longer than **v**, any values of TRUE in positions of **b** beyond the length of **v** will produce NA values in the result. To

Description	R expression	MATLAB expression
True where **v** is less than a	v < a	v < a
True where **v** is greater than a	v > a	v > a
True where **v** is less than or equal to a	v <= a	v <= a
True where **v** is greater than or equal to a	v >= a	v >= a
True where **v** is equal to a	v == a	v == a
True where **v** is not equal to a	v != a	v ~= a

TABLE 3.1
Comparison operators. These can be applied to scalars, vectors, or matrices.

understand why, note that v[b] is equivalent to v[which(b)] as long as **b** is at least as long as **v**. (If **b** is shorter, it is first recycled up to the length of **v**.)

(MATLAB): If the length n of **b** is shorter than **v**, only the first n elements of **v** are used. If **b** is longer than **v**, an error is given. To understand why, note that v(b) is equivalent to v(find(b)).

29. A vector containing only the positive elements of vector **v**

v[v > 0]	v(v > 0)

30. Given vector **b** containing logical values, and vector **w**, set the elements of **v** whose corresponding elements of **b** are TRUE to the consecutive elements of **w**. **w** may also be a scalar. See the following entry for a common example.

v[b]=w	v(b)=w

For example, say the second, fifth, and sixth entries of **b** are TRUE and the other entries are FALSE, and the vector **w** (of length 3) contains the values 10, 11, and 12. Then the second, fifth, and sixth elements of **v** will be set to 10, 11, and 12, and other elements of **v** will remain unchanged.

31. Set the negative entries of **v** to zero.

v[v < 0] = 0	v(v < 0) = 0

32. All but the kth element of vector **v**

v[-k]	v(k) = [ ] will do it, but modifies the original vector **v**. To do it without modifying **v**, you need to either build a vector containing all indices into **v** except for k, i.e., v([1:(k-1) (k+1):end]) or use logical vector indexing as in entry 28: v(~ismember(1:end,k))

33. All but the second, fourth, and seventh elements of **v**

v[-c(2,4,7)] or v[c(-2,-4,-7)]	v(~ismember(1:end,[2 4 7]))

34. All elements of **v** which are in vector **w**

v[!is.na(match(v,w))]	v(ismember(v,w))

35. All elements of **v** which are not in vector **w**

v[is.na(match(v,w))]	v(~ismember(v,w))

36. Truncate vector **v**, keeping only the first 10 elements

v = v[1:10] or length(v)=10	v = v(1:10)

37. For a vector **v** which has less than 10 elements, "grow" the vector to have length 10

length(v)=10 or v[10]=NA	v(10)=0

In both platforms, trying to store a value beyond the end of a vector simply grows the vector. Doing this repeatedly is deceptively inefficient, however. When you grow a vector in this way, a new block of memory is allocated for the new, larger vector, and the previous contents are copied over to the new memory. If you know your vector will repeatedly need to grow, it is best to allocate sufficient space for all of its values ahead of time to avoid repeated copying.

R pads vectors with NA when growing them; MATLAB pads them with zeros.

3.4 Creating matrices

38. Create a matrix containing arbitrary values: $\begin{bmatrix} 4 & 8 & 15 \\ 16 & 23 & 42 \end{bmatrix}$

R	MATLAB
matrix(c(4,8,15,16,23,42), nrow=2,byrow=TRUE) to enter the values by rows, or matrix(c(4,16,8,23,15,42), nrow=2) to enter them by columns. The parameter **ncol** can be used to specify the number of columns in the matrix, instead of or in addition to **nrow** to specify the number of rows.	[4 8 15 ; 16 23 42] or [4 8 15 16 23 42]

(R): Values are recycled here if you do not provide enough values when creating a matrix (see item 24). That is, matrix(1:3, nrow=2, ncol=6) creates the matrix

$$\begin{bmatrix} 1 & 3 & 2 & 1 & 3 & 2 \\ 2 & 1 & 3 & 2 & 1 & 3 \end{bmatrix}.$$

You can also construct matrices by reshaping vectors. That is, you can do this: A=c(4, 16, 8, 23, 15, 42); dim(A) = c(2,3). Note that the values must be provided in column order. If you wish to provide them in row order, you must transpose: A = c(4, 8, 15, 16, 23, 42); dim(A)=c(3,2); A = t(A).

39. $m \times n$ matrix of zeros

matrix(0,nrow=m,ncol=n) or simply matrix(0,m,n)	zeros(m,n) or zeros([m n])

(R): This is actually another case of recycling of values; the single value is recycled enough times to fill in the entire matrix.

(MATLAB): The latter form is more useful if you have the desired dimensions stored in a vector **v**, i.e., you can do `zeros(v)`.

40. $m \times n$ matrix containing the value j in all positions

`matrix(j,m,n)`	`j*ones(m,n)` or `repmat(j,m,n)`

(R): This is yet another case of recycling of values.

(MATLAB): Again, to facilitate using a vector containing the desired dimensions, you can also use `j*ones([m n])` or `repmat(j,[m n])`.

41. $n \times n$ identity matrix I_n

`diag(n)`	`eye(n)`

42. Build a diagonal matrix A using elements from vector **v**

`diag(v)` works, although if **v** is a vector of length 1 containing the value n, then this command will behave like the previous one and create an $n \times n$ identity matrix, rather than a 1×1 matrix containing the value n. To ensure this problem does not occur, you can instead use `diag(v,nrow=length(v))`.	`diag(v)`

43. Build a matrix by stacking matrix A and B on top of each other (they must have the same numbers of columns)

`rbind(A,B)`	`[A ; B]`

(MATLAB): This is actually the same syntax as used in items 1 and 38. The syntax works whether you use scalars as in that entry, or matrices, as long as the matrices' dimensions are conformable.

44. Build a matrix by gluing matrix A and B beside each other left-to-right (they must have the same numbers of rows)

`cbind(A,B)`	`[A B]`

(MATLAB) See the note following the previous entry.

45. Given vector **v**, create a matrix which has k columns, each of which is a copy of **v**

`matrix(rep(v,k),ncol=k)`	If **v** is a column vector, then `repmat(v,1,k)`. If **v** is a row vector, then transpose it, i.e., `repmat(v',1,k)`. If you are unsure if **v** will be a row or column vector, force it to be treated as a column as in entry 18, via `repmat(v(:),1,k)`. The indexing method from entry 60 can also be used; if **v** is a column vector, then `v(:,ones(1,k))`. If **v** is a row vector, then `v(ones(1,k),:)'`. This latter method is called "Tony's trick," but it may be a bit slower.

46. Given vector **v**, create a matrix which has k rows, each of which is a copy of **v**

Either simply transpose the previous item, i.e., `t(matrix(rep(v,k),ncol=k))`, or construct it directly via `matrix(rep(w,each=k),nrow=k)`	Transpose one of the methods from the previous item, e.g., `repmat(v(:),1,k)'`.

47. Combine the previous two items: given vectors **x** and **y** of lengths m and n, respectively, build $n \times m$ matrices **X** whose rows are copies of **x** and **Y** whose columns are copies of **y**

`m=length(x); n=length(y); X=` `matrix(rep(x,each=n),nrow=n);` `Y=matrix(rep(y,m),nrow=n)`	`[X,Y]=meshgrid(x,y)`

This operation is actually very useful, which is why a function is provided to do it in MATLAB. If you build such matrices **X** and **Y**, you can then use matrix computations to construct a matrix **Z** using a function $Z = f(X, Y)$ to plot as a contour plot or image/heatmap plot.

48. $n \times n$ Hilbert matrix H where $H_{ij} = 1/(i + j - 1)$

`Hilbert(n)`, but this is part of the **Matrix** package (see Section 13.9 for information about installing/loading packages).	`hilb(n)`

3.5 Working with matrices

One of the most noticeable differences between R and MATLAB is how you access an entire row or column of a matrix. In R, you simply leave out the row index in order to specify that you want all rows. In MATLAB, you must use the special symbol ":" as the row index in order to specify all rows. If you have gotten used to the MATLAB way of things, the R syntax looks quite strange, with commas that are not actually separating two symbols.

49. Access or set the element in row 2, column 3 of matrix A

R	MATLAB
`A[2,3]` or `A[2,3]=17`	`A(2,3)` or `A(2,3)=17`

See the note following entry 16 regarding parentheses and square brackets. See item 73 for what happens if you store a value outside the bounds of a matrix.

50. Use a single index to access or set an element of matrix A

`A[5]` or `A[5]=42`	`A(5)` or `A(5)=42`

When accessing matrix elements with a single index, the index counts down the first column, then down the second column, etc. So for a matrix with 5 rows, then `A[5]` or `A(5)` refers to the last element in the first column; `A[6]` or `A(6)` refers to the first element in the second column; and so on.

51. Use a set of single indices to access elements of matrix A

`A[c(5,8,7)]`	`A([5 8 7])` for result as a row vector, or `A([5;8;7])` for a column vector

52. Extract the diagonal elements of matrix A, as a vector

`diag(A)`	`diag(A)` Note: this produces a column vector

53. The dimensions of matrix **A** (i.e., numbers of rows and columns) as a vector

`dim(A)`	`size(A)`

54. Number of rows or columns in matrix A

`nrow(A)` or `ncol(A)`, or you could use `dim(A)[1]` or `dim(A)[2]`	`size(A,1)` or `size(A,2)`

(R): You may wish to use `NROW(A)` or `NCOL(A)` if **A** could possibly be a vector. In that case, the latter functions treat the vector **A** of length n as if it were a column vector, or more specifically an $n \times 1$ matrix. That is, `NROW(A)` and `NCOL(A)` return n and 1, respectively. `nrow(A)` and `ncol(A)` both return NULL if **A** is a vector rather than a matrix.

55. The maximum dimension of **A**, i.e., the larger of the numbers of rows and columns

`max(dim(A))`	`length(A)`

56. Column 2 of matrix **A**

`A[,2]` Note: this gives the result as a vector. To have the result instead be an $m \times 1$ matrix, use `A[,2,drop=FALSE]`. The **drop=FALSE** argument indicates that you do not want to "drop" dimensions that have length 1.	`A(:,2)`

This is a rather glaring difference in syntax between the two packages. In MATLAB, you use the special index ":" to indicate that you want all elements along a given dimension. In R, you simply leave out the indices for that dimension, which may look somewhat strange until you get used to it. To further demonstrate, in R, `A[,]` is equivalent to `A`, i.e., it gives you the entire matrix. In MATLAB, the equivalent is `A(:,:)`.

57. Row 7 of matrix **A**

`A[7,]` gives the result as a vector. As in the previous entry, use `A[7,,drop=FALSE]` to give the result as a $1 \times n$ matrix.	`A(7,:)`

58. All elements of **A** (column by column) as a single vector

`c(A)`	`A(:)` (this gives a column vector)

59. Rows 2–4, columns 6–10 of **A** (this is a 3×5 matrix)

`A[2:4,6:10]`	`A(2:4,6:10)`

As the next entry points out, you do not need to use a set of consecutive values when selecting the rows or columns, and in fact you can choose a given row or column more than once.

60. A 3×2 matrix consisting of rows 7, 7, and 6 and columns 2 and 1 of A (in that order)

`A[c(7,7,6),c(2,1)]`	`A([7 7 6], [2 1])`

When using vectors **r** (of length p) and **c** (of length q) as the row and column indices with a matrix **A**, all columns are used in combination with each row, i.e., the result is a $p \times q$ matrix. This idea can be used to set all elements within a specified submatrix as well (see entry 65).

61. Given a single index **ind** into matrix **A** (as in entry 50), compute the row **r** and column **c** of that position.

`tmp = arrayInd(ind, dim(A))` `r = tmp[,1]; c = tmp[,2]`	`[r,c] = ind2sub(size(A), ind)`

The commands above work even if **ind** is a vector containing many such indices; in that case, **r** and **c** will be corresponding vectors of the same size.

62. Given the row **r** and column **c** of an element in matrix **A**, compute the single index **ind** which can be used to access that element (as in entry 50).

`ind = (c-1)*nrow(A)+r`	`ind = sub2ind(size(A), r, c)`

The commands above work even if **r** and **c** are vectors containing many such indices; in that case, **ind** will be a corresponding vector of the same size.

63. Given equal-sized vectors **ind** and **v**, use single-indexing (as in entry 50) so that the kth value in **v** is stored in the element of **A** referred to by the kth element of **ind**, i.e., set many specified elements of the matrix **A** at once. If a scalar value **v** is provided instead, all specified elements of **A** will be set to that value.

`A[ind]=v`	`A(ind)=v`

64. Given equal-sized vectors **r**, **c**, and **v**, store the kth value in **v** in the element of **A** which is in the row and column referred to by the kth elements of **r** and **c**, i.e., set many specified elements of the matrix **A** at once. (A scalar value **v** can also be provided instead of a vector.)

`A[cbind(r,c)]=v` will work, as will `A[(c-1)*nrow(A)+r] = v`	`A(sub2ind(size(A),r,c))=v`

(R): The first method utilizes the fact that if you use a matrix **B** which has two columns to index **A**, i.e., **A**[**B**], then R interprets the two columns of **B** as row and column indices into **A**. The second method simply uses single-indexing as computed in entry 62.

65. Given vectors **r** (of length p) and **c** (of length q) and $p \times q$ matrix **B**, set the submatrix of **A** using row and column indices from **r** and **c** equal to **B** (see entry 60).[1]

`A[r,c]=B`	`A(r,c)=B`

A scalar value can be provided, rather than matrix **B**, in which case all elements of the specified submatrix of **A** will be set to that value.

---

[1] I will leave it as an exercise for the reader to determine what happens if either **r** or **c** has repeated indices as in entry 62, which would then try to store multiple elements of **B** into the same positions of **A**.

66. The transpose of the matrix **A**.

t(A)	A.'

67. The complex conjugate transpose of the matrix **A**.

Conj(t(A))	A'

(MATLAB): If **A** only contains real values, then `A'` and `A.'` give the same result; in that case, the former is frequently used for simplicity.

68. Circularly shift the rows of matrix A down by s_1 elements, and right by s_2 elements

Modulo arithmetic on the row and column indices can be used:
`A[(1:nrow(A)-s1-1)%%nrow(A)+1,`
`(1:ncol(A)-s2-1)%%ncol(A)+1]`
	circshift(A, [s1 s2])

69. Flip the order of elements in each row of matrix A

t(apply(A,1,rev)) or A[,ncol(A):1]	fliplr(A)

(R): The **apply** function returns its results down the columns of a matrix (even if you are operating on rows); this is why it is necessary to transpose the result. See Section 4.1 for more information.

70. Flip the order of elements in each column of matrix A

apply(A,2,rev) or A[nrow(A):1,]	flipud(A)

71. Given matrix **A**, create a new matrix **L** containing the lower-triangular portion of **A** (i.e., with all elements above the diagonal set to zero).

L=A; L[upper.tri(L)]=0	L = tril(A)

(R): See the logical indexing entries below, in particular entry 80, for how this and the following entry work.

72. Given matrix **A**, create a new matrix **U** containing the upper-triangular portion of **A** (i.e., with all elements below the diagonal set to zero).

U=A; U[lower.tri(U)]=0	U = triu(A)

73. For a matrix **A** which has dimensions less than 6×8, "grow" the matrix to have those dimensions.

tmp = matrix(NA,6,8) tmp[1:nrow(A),1:ncol(A)] = A A = tmp	A(6,8)=0

(R): Note that R will grow a vector if you store a value off its end, but it will not do the same with a matrix (instead it produces an error). There is no simple way to expand a matrix. The commands above first produce an appropriately sized matrix of NAs in **tmp**, copies the original matrix **A** into part of it, and then stores the results back in **A**. The value

NA was used here for consistency with growing vectors as described in entry 37, but any value could be used.

$$* * * * *$$

Note that the various commands from entries 26–31 using logical values with vectors also work with matrices. Essentially, the matrices work like vectors when doing so, by laying out their elements column by column. Single-indexing (as in entry 50) is also used in these commands.

74. Given a matrix **B** containing logical values, create a vector giving the (single-) indices of the TRUE elements.

`which(B)`	`find(B)`

75. The (single-) indices of the positive elements of matrix **A**

`which(A > 0)`	`find(A > 0)`

76. Build vectors **r** and **c** containing the rows and columns of the positive elements of **A**.

`tmp = which(A > 0, arr.ind=TRUE);` `r=tmp[,1]; c=tmp[,2]`	`[r,c] = find(A > 0)`

77. Given a vector **b** containing logical (TRUE/FALSE) values, build a vector by extracting only those elements of matrix **A** whose corresponding elements of **b** are TRUE

`A[b]`	`A(b)`

For both platforms, if you use a matrix **B** rather than a vector **b**, the results are still returned as a vector.

78. A vector containing only the positive elements of matrix **A**

`A[A > 0]`	`A(A > 0)`

79. Given vector **b** containing logical values, and vector **w**, set the elements of **A** whose corresponding elements of **b** are TRUE to the consecutive elements of **w**. **w** may also be a scalar. See the following entry for a common example.

`A[b]=w`	`A(b)=w`

80. Set the negative entries of **A** to zero.

`A[A < 0] = 0`	`A(A < 0) = 0`

3.6 Reshaping matrices, and higher-dimensional arrays

It is possible to create higher-dimensional arrays in both platforms, i.e., arrays that have more than just the two dimensions of rows and columns. Note that in R, an **array** is a data type that can have any number of dimensions; a **matrix** is a special case of an array that has exactly two dimensions. In MATLAB, there is no fundamental difference between an array with two dimensions and an array with more than two dimensions.

In both platforms, the easiest way to construct an array with more than two dimensions is either to take a vector containing the desired values and turn it into an array with the desired dimensions, or to create an "empty" array (one containing zeros) and then fill in its entries one two-dimensional "slice" at a time.

81. Reshape a 2-D matrix/array to have m rows and n columns.

R	MATLAB
`dim(A) = c(m,n)`	`A = reshape(A,m,n)`

82. Number of dimensions of array **A**.

`length(dim(A))`	`ndims(A)`

83. The actual dimensions of array **A** (the dimensions are returned in a vector).

`dim(A)`	`size(A)`

84. Total number of elements in array **A**.

`length(A)`	`numel(A)`

85. Create a $3 \times 4 \times 2$ array containing all zeros.

`A=array(0,c(3,4,2))`	`A=zeros(3,4,2)` or `A=repmat(0,[3 4 2])`

(MATLAB): `zeros([3 4 2])` also works, but note that `repmat(0,3,4,2)` does not.

86. Create a $3 \times 4 \times 2$ array containing the values from the vector **v**. Values are entered in the array with the earlier dimensions varying most rapidly. That is, if the three dimensions are referred to as rows, columns, and depths, the elements of **v** are placed down the rows of the first column and first depth, then down the rows of the second column and first depth, and so on until the rows of all columns of the first depth are filled. Subsequent elements are then placed into the second depth.

`A=array(v,c(3,4,2))`	`A=reshape(v,3,4,2)` or `A=reshape(v,[3 4 2])`

(R): Values in **v** are recycled if needed (see entry 24).

(MATLAB): The vector **v** must have exactly 24 elements in it for this example, otherwise an error is produced. In general, `A=reshape(v,dims)` will fail if **length(v)** does not equal **prod(dims)**.

87. Given array **A**, change its dimensions to be $2 \times 6 \times 2$.

`dim(A)=c(2,6,2)`	`A=reshape(A,2,6,2)` or `A=reshape(A,[2 6 2])`

In either package, an error occurs if the number of elements in **A** does not match the product of the dimensions specified.

88. Given an array of dimensions $d_1 \times d_2 \times d_3$, extract a two-dimensional $d_1 \times d_2$ "slice" (with third index of **i3**).

To extract as a $d_1 \times d_2$ matrix: `A[,,i3]`. To extract as a $d_1 \times d_2 \times 1$ 3-D array: `A[,,i3,drop=FALSE]`	To extract as a $d_1 \times d_2 \times 1$ 3-D array: `A(:,:,i3)`. To extract as a $d_1 \times d_2$ array: `squeeze(A(:,:,i3))`

(MATLAB): The **squeeze** function will have no effect on 2-D arrays; this is so that `squeeze(r)` where **r** is a row vector will remain a row vector, rather than dropping the first dimension which has length 1 and turning it into a column vector.

89. Given an array of dimensions $d_1 \times d_2 \times d_3 \times d_4$, set a two-dimensional $d_1 \times d_2$ "slice" (with third and fourth indices of **i3** and **i4**, respectively) equal to the matrix **B**.

`A[,,i3,i4]=B`	`A(:,:,i3,i4)=B`

3.7 Sparse matrices

Both platforms have facilities for working with sparse matrices, that is, matrices which are (typically) large and which contain mostly zeros. Such matrices have a more compact internal representation than standard matrices. Once they have been constructed, standard matrix operations may then be used on sparse matrices.

In R, the **Matrix** package allows you to construct sparse matrices using either the **Matrix** function (to construct a sparse matrix from a dense one) or the **sparseMatrix** function.

In MATLAB, the **sparse** function can be used to construct a sparse matrix. Other useful functions are **sprand**, **spconvert**, **spfun**, and **nnz**.

3.8 Names with vectors and matrices/arrays

Both platforms provide ways of associating names with vectors, matrices, or arrays, allowing one to access elements by name rather than simply by number. However, the mechanisms are quite different between the two platforms.

3.8.1 R: names for vector/matrix elements and matrix rows/columns

R allows you to give names to the elements of a vector, and to the rows and/or columns of a matrix if desired. Those names are displayed when the vector or matrix is displayed, and can also be used to access their elements. The **names** function can be used to get or set the names of the elements of a vector, while **colnames** and **rownames** can be used with a matrix. Note that the latter two are really special cases of working with **dimnames** to access a list of the dimension names. The following commands demonstrate names associated with a vector.

```
R
> v = c(4,8,15)
> v
[1]  4  8 15
```

```
> names(v) = c('one','hmmm','OK')
> v
 one hmmm   OK
   4    8   15
> v[1]
one
  4
> v['one']
one
  4
> names(v)[1] = 'new'
> v['new'] = 16
> v
 new hmmm   OK
  16    8   15
> w=c(foo=4, bar=8, baz=15)
> w
foo bar baz
  4   8  15
```

The following commands demonstrate this feature for a matrix:

```
————————————————— R ———————————————
> A = matrix(c(4,8,15,16,23,42),nrow=2,byrow=TRUE)
> A
     [,1] [,2] [,3]
[1,]    4    8   15
[2,]   16   23   42
> colnames(A) = c('one','hmmm','OK')
> rownames(A) = c('first', 'second')
> A
       one hmmm OK
first    4    8 15
second  16   23 42
> A['second','hmmm']
[1] 23
> cn = c('fee', 'fi', 'fo')
> rn = c('hi', 'lo')
> dimnames(A) = list(rn,cn)
> A
   fee fi fo
hi   4  8 15
lo  16 23 42
> B=matrix(1:6,nrow=2,dimnames=list(rn,cn))
> B
   fee fi fo
hi   1  3  5
lo   2  4  6
```

You can also assign names to individual elements of a matrix. Doing so, and working with those names, is done in the same way as it is for vectors. For example, names(A)=c('a', 'b', 'c', 'd', 'e', 'f'); A['b'] = 50.

3.8.2 R data frames

A data frame is a matrix in which the different columns (referred to as components) can contain different types of data, such as numeric, logical, factor, character, etc. Data frames are very commonly used in R for data analysis and statistics.

```
──────────────────────── R ────────────────────────
> color = c('red','red','blue','red','blue','blue')
> size = c('M','XL','M','XL','L','XL')
> size = factor(size, ordered=TRUE, levels=c('M','L','XL'))
> IDNum = c(4, 8, 15, 16, 23, 42)
> quality = c(42.1, 17.2, 121.3, 7.4, 11.5, 55.6)
> d = data.frame(IDNum, color, size, quality)
> d
  IDNum color size quality
1     4   red    M    42.1
2     8   red   XL    17.2
3    15  blue    M   121.3
4    16   red   XL     7.4
5    23  blue    L    11.5
6    42  blue   XL    55.6
```

You can see above that R creates column labels from the names of the variables provided to **data.frame**, and also applies the default row labels 1 through 6. You can provide alternate column names via tags:

```
──────────────────────── R ────────────────────────
> d = data.frame(ID=IDNum, hue=color, scale=size, qual=quality)
> d
  ID  hue scale  qual
1  4  red     M  42.1
2  8  red    XL  17.2
3 15 blue     M 121.3
4 16  red    XL   7.4
5 23 blue     L  11.5
6 42 blue    XL  55.6
```

You can specify different row names as well. You can provide a vector with the names (a numeric vector will be converted to character strings, and the vector you provide here can be the same as one of the data frame's columns). Alternatively, you can specify that one of the data frame's columns should instead be used as the labels (note that this removes it as one of the regular columns); this can be done either by giving a column number or name.

```
──────────────────────── R ────────────────────────
> d = data.frame(ID=IDNum, hue=color, scale=size, qual=quality,
+ row.names=quality)
> d
        ID  hue scale  qual
42.1     4  red     M  42.1
17.2     8  red    XL  17.2
121.3   15 blue     M 121.3
7.4     16  red    XL   7.4
11.5    23 blue     L  11.5
55.6    42 blue    XL  55.6
```

```
> d = data.frame(ID=IDNum, hue=color, scale=size, qual=quality,
+ row.names='qual')   # Note: same results achieved using row.names=4
> d
        ID  hue scale
42.1    4   red    M
17.2    8   red    XL
121.3  15  blue    M
7.4    16   red    XL
11.5   23  blue    L
55.6   42  blue    XL
```

In the last example above, the same results could have been achieved using **row.names=4**, because **qual** was the fourth column in the data frame.

You can access values within a data frame in a couple of different ways. First, you can use matrix-like indexing, to access single elements, or slices across rows or down columns. For example, using the final data frame **d** defined above:

```
——————————————————————— R ———————————————————————
> d[4,1]
[1] 16
> d[,1]
[1]  4  8 15 16 23 42
> d[4,]
     ID hue scale
7.4 16 red    XL
```

Note that the row name **7.4** appears when we select all of row 4 in the final command above. There are two more ways to extract an entire column from a data frame: using single-indexing, or via dollar-sign matching:

```
——————————————————————— R ———————————————————————
> d[1]
        ID
42.1    4
17.2    8
121.3  15
7.4    16
11.5   23
55.6   42
> d$ID
[1]  4  8 15 16 23 42
```

Single-indexing extracts the column still inside a data frame (note that multiple columns could be extracted in this way, e.g., d[c(1,3)]), while using dollar-sign matching extracts the data as a standard vector variable (as a numeric vector in the example above).

Because data are so often stored in data frames for analysis, and it can become cumbersome to repeatedly use notation like **d$ID** to access its components, there are a couple of convenient ways to work with a particular data frame: via **attach/detach** and via **with**.

If you have a data frame named **d**, the command **attach(d)** will add the data frame to R's search path. This means that you can simply type **hue** rather than **d$hue** to access that component of the data frame. For example, you can enter **str(hue)** to see the structure of that component. You can then enter **detach(d)** to remove **d** from the search path. There are a couple of caveats to note about the use of **attach**: first, say you already have a variable not within the data frame named, for example **ID**. This variable will mask the **ID** component within the data frame when you attach it (by default, **attach** will display a message warning you of this). You can still use the explicit **d\$ID** notation to access that component of the data frame. Second, if you try to make a change to a component of the data frame, e.g., via a command like `scale = rep('L', 6)`, it does not change that component of the data frame. Instead, it creates a new variable named **scale** within your user workspace and assigns the value there; this new variable will still exist after you **detach** your data frame. Technically, only a copy of the data frame is added to the new environment constructed by **attach**, so even other tricks such as calling **assign** will not assign values to the data frame itself; you must explicitly refer to the data frame (e.g., `d$scale = rep('L', 6)` to assign values to its components. Note that in addition to data frames, you can also attach lists.

A second method provided to more conveniently work with data frames (and lists) is the **with** function. You can use **with(d, expr)** to evaluate the given expression **expr** with **d** attached. Technically, a temporary environment is constructed containing the components of **d** available, and the expression evaluated within that environment. If you want to evaluate multiple commands inside the **with**, enclose them with curly brackets:

```R
with(d, {print(summary(hue)); plot(scale)})
```

Note that you must use **print** to display the summary of **hue**, just as you would within a script or function (see Page 79). Also, because the commands in the **with** statement are executed within a temporary environment, any variables created or modified are limited to that environment, just as they would be within a function. To construct a new variable or make changes to an existing one that persist beyond the call to **with**, use the **<<-** operator (see Page 87). If you wish to modify your data frame, you can use **within** rather than **with**. Any changes made to the data frame inside a **within** statement will be stored within a local copy of the data frame which is then returned by **within**, as in the following example (you must then store this copy of the data frame, back in the original or somewhere else if you wish to keep it):

```R
> new.d = within(d, {ID=6:1})
> new.d
        ID  hue scale
42.1     6  red    M
17.2     5  red   XL
121.3    4 blue    M
7.4      3  red   XL
11.5     2 blue    L
55.6     1 blue   XL
```

3.8.3 MATLAB structs

MATLAB has a data type called a "structure." A structure can contain values of different types, which can be referred to by name. To understand the behavior of structures, you should remember that like other MATLAB data types, structures are fundamentally arrays. You can create a structure via the **struct** command.

```MATLAB
>> s = struct('ID', {4 8 15 16}, 'hue', {'red','red','blue','red'}, ...
   'scale', {'M', 'XL', 'M', 'XL'}, 'qual', {42.1, 17.2, 121.3, 7.4})
s =
```

```
1x4 struct array with fields:
    ID
    hue
    scale
    qual
>> s.ID
ans =
    4
ans =
    8
ans =
    15
ans =
    16
>> s(3)
ans =
       ID: 15
      hue: 'blue'
    scale: 'M'
     qual: 121.3000
>> x = s.ID
x =
    4
>> s.ID(3)
Field reference for multiple structure elements that is followed by more
    reference blocks is an error.
>> s(3).ID
ans =
    15
```

The above command creates a structure with four fields named ID, hue, scale, and qual. Four corresponding 1×4 cell arrays are also passed to **struct** (cell arrays are explained in Chapter 5). The resulting structure s is also 1×4. You can use the syntax **s.ID** to refer to the ID field of the structure s; note that you get back the four values as four different return values. If you examine **s(3)**, you get all fields from the third element of the structure. To access the third ID, you must access the ID field of the third element of the structure via **s(3).ID**. The command **s.ID(3)** fails, because s is a 1×4 vector; the **ID** field of s is not.

If you really wish to create a 1×1 structure whose fields contain arrays, then you can pass in 1×1 cell arrays as the corresponding values. The elements of those cell arrays can be vectors, as below.

```
——————————————— MATLAB ———————————————
>> s = struct('ID', {[4 8 15 16]}, 'hue', {{'red','red','blue','red'}}, ...
'scale', {{'M', 'XL', 'M', 'XL'}}, 'qual', {[42.1, 17.2, 121.3, 7.4]})
s =
       ID: [4 8 15 16]
      hue: {'red'  'red'  'blue'  'red'}
    scale: {'M'    'XL'    'M'    'XL'}
     qual: [42.1000 17.2000 121.3000 7.4000]
>> s.ID
ans =
    4      8      15      16
```

```
>> s.ID(3)
ans =
     15
```

Note that the strings passed in for the hue and scale fields still need to be within cell arrays, to avoid the strings being concatenated into the combined strings "redredbluered" and "MXLMXL" which would happen if regular vectors were used.

3.9   Miscellaneous

In R, you can use subscripting on expressions; e.g., `f(A)[1,2]` to access the value in row 1, column 2 of the matrix returned by `f(A)`. In MATLAB, one way to do this is to use a temporary variable, e.g., `tmp=f(A); tmp(1,2)`. Another way to do it is with the help of the **subsref** function, which is what handles array referencing: `S.type='()'; S.subs={1,2}; subsref(f(A),S)`. This latter trick can be accomplished without use of a temporary variable by instead building a temporary structure using **struct**: `subsref(f(A), struct('type', '()', 'subs', {{1,2}}))`.

4

Matrix/Vector Calculations and Functions

Both platforms are designed to work natively with vectors and matrices. Many of the standard functions to perform basic mathematical calculations are *vectorized*, meaning they will operate on all elements of a vector or matrix. That is, you can write things like sqrt(v) to compute the square roots of all elements of vector **v**, or exp(A) to raise e to the power of each element of matrix **A**. In a non-vectorized platform like the C programming language, a **for** loop would be needed to iterate over the elements of the vector or matrix. In both R and MATLAB, that **for** loop is internal, running at the speed of code compiled to the native computer language, and therefore runs much more quickly than a loop you would write yourself, which runs more slowly due to R and MATLAB being interpreted languages.[1]

4.1 Applying a function to rows or columns of a matrix

One very useful operation is to apply a function to parts of a matrix. It can be useful to do this along a certain dimension of a matrix. For example, you may wish to compute the means of all columns or rows of a matrix. Because those particular examples are so commonly needed, there are already facilities to compute them (see Section 4.4). But suppose you wish to compute the sum of the cubes of the elements in the rows or columns of a matrix.

R

R has the very useful function **apply** which lets you apply a function along particular dimensions of a matrix. To apply a function **f** (which should receive a vector as its parameter) to the rows of a matrix **A**, use **apply(A, 1, f)**. To apply the function to the columns of **A**, use **apply(A, 2, f)**. The second parameter to **apply** specifies the "margins" or dimensions that you want to keep in the results; keeping margin m means you apply the function to all elements whose m^{th} index is 1, then to all elements whose m^{th} index is 2, and so on. If the function you are applying to a matrix returns scalars, then **apply** will return a vector containing the resulting values. If the function you are applying returns a vector which is always the same length, then **apply** returns a matrix whose k^{th} column has the results of applying the function to the kth slice of the matrix (regardless of which margin you are applying it to, i.e., rows or columns). If the function you are applying returns vectors which are of different lengths for the different slices of your matrix, then **apply** returns its results in a list (see Chapter 5 for more about "Lists").

The following code performs the computation described in the example above, summing the cubes of the elements of the rows or columns of **A**.

─────────────────────────── R ───────────────────────────
```
> a=matrix(1:12,nrow=3,byrow=TRUE)
```

───
[1]Recent versions of MATLAB are fairly aggressive about optimizing simple **for** loops, so they often run much more quickly than analagous loops in R.

```
> sumcubes=function(v) { return(sum(v^3)) }  # sums cubes of elements of v
> apply(a, 1, sumcubes)  # sums of the rows
[1]   100 1196 4788
> apply(a, 2, sumcubes)  # sums of the columns
[1]   855 1224 1701 2304
```

See Section 8.1.1 for more information about writing functions.

MATLAB

MATLAB does not have a function directly equivalent to R's **apply**, but there is a way
to achieve the same effect. First, the matrix **A** can be turned into a cell array, with the
elements of the cell array containing either the rows or columns of **A**. The **cellfun** function
can then be used to apply a given function to that cell array, i.e., the rows or columns of
the original matrix.

```
───────────────── MATLAB ─────────────────
>> A = [1 2 3 4 ; 5 6 7 8; 9 10 11 12];
>> numRows = size(A,1);  numCols = size(A,2);  % for convenience
>> % rowsCell contains the rows of A in a 3x1 cell array
>> rowsCell = mat2cell(A, ones(1,numRows), numCols)
rowsCell =
    [1x4 double]
    [1x4 double]
    [1x4 double]
>> % colsCell contains the columns of A in a 1x4 cell array
>> colsCell = mat2cell(A, numRows, ones(1, numCols))
colsCell =
    [3x1 double]    [3x1 double]    [3x1 double]    [3x1 double]
>> sumcubes = @(v) sum(v.^3);  % sums cubes of elements of v
>> cellfun(sumcubes,rowsCell)  % sums of the rows
ans =
         100
        1196
        4788
>> cellfun(sumcubes,colsCell)  % sums of the columns
ans =
        855        1224        1701        2304
```

See Section 8.2.1 for information about writing anonymous functions like **sumcubes** above,
and Section 8.2.7 for information about creating handles to built-in functions which can then
be passed as parameters to **cellfun**.

4.2   Applying a function to all elements of a matrix

You may also wish to apply a function to every element of a matrix. For example, if **A** is
an $m \times n$ matrix containing integers, we may wish to generate a new matrix **B**, where the
number b_{ij} in row i, column j is a random integer from within the range 1 to a_{ij} inclusive.

R

The **apply** function can work here as well. We first write a function **myfunc(n)** which generates a single random integer within the range 1 to **n**, and then **apply** that function to all elements of our matrix **A**:

```R
> A=matrix(1:12,nrow=3,byrow=TRUE)
> myfunc = function(n) { return(sample(n, 1)) }
> apply(A, c(1,2), myfunc) #gives different results when called repeatedly
     [,1] [,2] [,3] [,4]
[1,]   1    1    3    3
[2,]   4    5    6    5
[3,]   1    2    5   11
```

MATLAB

The **arrayfun** function gives the desired behavior. We can use the **randi** function, since **randi(n)** will generate a single integer from 1 to **n**, and apply that function to every element of **A**:

```MATLAB
>> A = [1 2 3 4 ; 5 6 7 8; 9 10 11 12];
>> arrayfun(@randi, A)  % gives different results when called repeatedly
ans =
     1    2    2    3
     1    4    5    3
     3    2    8    2
```

4.3 Linear algebra calculations with vectors and matrices

1. Compute the dot product of vectors **x** and **y**.

R	MATLAB
sum(x*y)	dot(x,y)

2. Compute the vector cross-product of **x** and **y**.

Not in base R, but you can use cross(x,y) after loading the **pracma** package.	cross(x,y)

3. Given matrices **A** and **B**, compute the matrix product AB.

A %*% B	A * B

4. Element-by-element multiplication of matrices **A** and **B**.

A * B	A .* B

5. Solve the matrix equation $A\vec{x} = \vec{b}$.

`solve(A,b)`	`A \ b`

(R): This only works with square invertible matrices.

(MATLAB): If **A** is square and singular (or non-square and has rank less than the number of rows), you will receive a warning. If the system is inconsistent, MATLAB will compute a least-squares solution, i.e., a vector $\vec{y}$ which minimizes $A\vec{y} - \vec{b}$. Be especially aware of the fact that if **A** has fewer columns than rows, and $\vec{b}$ is not in the column space of **A**, MATLAB will give you a least-squares solution without any warnings or indication that it is not an actual solution.

6. Reduced echelon form of matrix **A**.

Not in base R, but the **rref** function from the **pracma** package will do it.	`rref(A)`

7. Determinant of matrix **A**.

`det(A)`	`det(A)`

8. Inverse of matrix **A**.

`solve(A)`	`inv(A)`

9. Trace of matrix **A**.

`sum(diag(A))`	`trace(A)`

10. Compute the matrix product AB^{-1}, assuming matrix **B** is invertible.

`A %*% solve(B)`	`A / B`

11. Compute the matrix product $A^{-1}B$, assuming matrix **A** is invertible.

`solve(A) %*% B`	`A \ B`

12. Element-by-element division of matrices **A** and **B**.

`A / B`	`A ./ B`

13. Raise matrix **A** to the kth power.

First load the **expm** package; then use `A %^% k`	`A ^ k`

14. Raise each element of matrix **A** to the kth power.

`A ^ k`	`A .^ k`

15. Rank of matrix **A**.

`qr(A)$rank`	`rank(A)`

16. Let **w** be a vector containing the eigenvalues of **A**, and **V** a matrix containing the corresponding eigenvectors.

`tmp = eigen(A); w=tmp$values; V=tmp$vectors`	`[V,D] = eig(A); w = diag(D)`

R's **eigen** function returns the eigenvalues and eigenvectors in a list, while MATLAB's **eig** function returns the eigenvalues packed within the diagonal elements of a matrix.

17. Perform an LU factorization of matrix $\mathbf{A}$, by finding matrices P, L, and U satisfying $PA = LU$.

`tmp = expand(lu(Matrix(A)));`
`L=tmp$L; U=tmp$U; P=solve(tmp$P)`
Note that **expand** and **lu** are part of the **Matrix** package.
	`[L,U,P] = lu(A)`

(R): The **lu** function does the factorization as $A = PLU$ rather than $PA = LU$, so if you really want it in the latter form, you need to invert the matrix P returned by **lu**, as in the commands above. The results will all be of class **Matrix**; if you want standard R matrices, you can convert via the commands `L = as.matrix(tmp$L); U = as.matrix(tmp$U); P = as.matrix(solve(tmp$P))*1`. Also note that LU factorization of dense matrices will work for square or non-square matrices, but sparse matrices must be square.

18. Singular-value decomposition: given an $m \times n$ matrix A, assuming $k = \min(m, n)$, find $m \times k$ matrix P with orthonal columns, square diagonal $k \times k$ matrix S, and $n \times k$ matrix Q with orthonormal columns such that $PSQ^T = A$.

`tmp=svd(A); P=tmp$u; S=diag(tmp$d);` `Q=tmp$v`	`[P,S,Q] = svd(A, 'econ')`

(MATLAB): If you omit the `'econ'` parameter, then S will have the same dimensions as A, P will be $m \times m$, Q will be $n \times n$, and P and Q will both be unitary matrices.

19. Schur decomposition of square matrix, $A = QTQ^*$ where T is upper triangular and Q is unitary, i.e., $Q^*Q = I$, with $Q^* = \overline{Q^T}$ the Hermitian (conjugate) transpose of Q.

`tmp=Schur(Matrix(A));`
`T=tmp@T; Q=tmp@Q`
Note that **Schur** is part of the Matrix package
	`[Q,T]=schur(A)`

(R): **T** and **Q** will be of class Matrix; if you want to make them into standard R matrices, instead do `T=as.matrix(tmp@T)` and `Q=as.matrix(tmp@Q)` above.

20. Cholesky factorization of a square, symmetric, positive definite matrix $A = R^*R$, where R is upper-triangular.

`R = chol(A)`	`R = chol(A)`

21. Permuted $AE = QR$ factorization of matrix A, where E is a permutation matrix, Q is orthogonal, and R is upper-triangular.

`tmp=qr(A); Q=qr.Q(tmp);`
`R=qr.R(tmp);`
`E=diag(ncol(A))[,tmp$pivot]`
	`[Q,R,E] = qr(A)`

22. Vector norms of vector $\mathbf{v}$ in $\mathbb{R}^n$: $\|\vec{v}\|_1 = \sum_{i=1}^{n} |v_i|$, $\|\vec{v}\|_2 = \left(\sum_{i=1}^{n} |v_i|^2\right)^{0.5}$, $\|\vec{v}\|_p = \left(\sum_{i=1}^{n} |v_i|^p\right)^{(1/p)}$, $\|\vec{v}\|_\infty = \max_{i=1}^{n} |v_i|$.

`N1=norm(matrix(v),'1');`
`N2=sqrt(sum(abs(v)^2));`
`Np=sum(abs(v)^p)^(1/p);`
`Ninf=norm(matrix(v),'i')`
	`N1=norm(v,1); N2=norm(v,2);`
	`Np=norm(v,p); Ninf=norm(v,inf)`

23. Matrix norms $\|A\|_1$, $\|A\|_2$, $\|A\|_\infty$, and Frobenius norm $\left(\sum_i (A^T A)_{ii}\right)^{1/2}$ of square matrix **A**.

`N1=norm(A,'1'); N2=norm(A,'2');`
`Ninf=norm(A,'i'); Nfrob=norm(A,'f')`
	`N1=norm(A,1); N2=norm(A);`
	`Ninf=norm(A,inf);`
	`Nfrob=norm(A,'fro')`

24. Condition numbers $\mathrm{cond}_1(A) = \|A\|_1\|A^{-1}\|_1$, $\mathrm{cond}_2(A) = \|A\|_2\|A^{-1}\|_2$, $\mathrm{cond}_\infty(A) = \|A\|_\infty\|A^{-1}\|_\infty$ of square matrix **A**.

`c1=1/rcond(A,'1');`
`c2=kappa(A,exact=TRUE);`
`cinf=1/rcond(A,'I')`
	`c1=cond(A,1); c2=cond(A,2);`
	`cinf=cond(A,inf)`

(MATLAB): There is also a function **rcond(A)** which computes the reciprocal of the condition number.

25. Orthonormal basis for the null space of matrix A.

`null(A)` This function is in package **pracma**.	`null(A)`

26. Orthonormal basis for the image/range/column space of matrix A.

`orth(A)` This function is in package **pracma**.	`orth(A)`

4.4   Statistical calculations

27. Mean of all elements in vector **v** or matrix **A**.

R	MATLAB
`mean(v)` and `mean(A)`	`mean(v)` and `mean(A(:))`

28. Means of all columns of matrix **A**.

`colMeans(A)`	`mean(A)`

29. Means of all rows of matrix **A**.

`rowMeans(A)`	`mean(A,2)`

30. Standard deviation of all elements in vector **v** or matrix **A**.

`sd(v)` and `sd(A)`	`std(v)` and `std(A(:))`

Both platforms use $n - 1$ as the denominator in the calculations, where n is the number of observations. In MATLAB, use `std(v,1)` or `std(A(:),1)` to instead use n as the denominator.

31. Standard deviation of columns of matrix **A**.

apply(A,2,sd)	std(A) or std(A,0,1) to normalize by $n-1$, or std(A,1,1) to normalize by n

An easy source of confusion is the difference in how you specify which dimensions to operate over in R vs. MATLAB. R's **apply** function takes the margins to retain as its second parameter; giving the value 2 as the margin above means you wish to retain the second dimension (the columns) in the result, that is, compute the standard deviation of each column. MATLAB's **std** function takes the dimension to compute the standard deviation along as its third parameter; giving the value 1 above means you compute the standard deviation along dimension 1 (the rows), and therefore perform the computation separately for each column.

32. Standard deviation of rows of matrix **A**.

apply(A,1,sd)	std(A,0,2) to normalize by $n-1$, or std(A,1,2) to normalize by n

33. Variance of all elements in vector **v** or matrix **A**.

var(v) and var(c(A))	var(v) and var(A(:))

As with standard deviations, both platforms use $n-1$ as the denominator in the calculations, where n is the number of observations. In MATLAB, use `var(v,1)` or `var(A(:),1)` to instead use n as the denominator.

34. Variance of columns of matrix **A**.

apply(A,2,var)	var(A) or var(A,0,1) to normalize by $n-1$, or var(A,1,1) to normalize by n

35. Variance of rows of matrix **A**.

apply(A,1,var)	var(A,0,2) to normalize by $n-1$, or var(A,1,2) to normalize by n

36. Mode of values in vector **v**.

There is no simple built-in function to compute the mode. The **table** or **unique** functions can be used to first generate a set of the unique values in **v**, and then used to extract the most frequent value. To choose the smallest value in case of a tie, use: `tmp=table(v); as.numeric(names(sort(-tmp)))[1]` To generate a vector of all values tied as most frequent, use: `tmp=table(v); as.numeric(names(tmp)[tmp == max(sort(tmp))])` To choose whichever of the tied values occurs first in **v**, use: `tmp=unique(v); tmp[which.max(tabulate(match(v, tmp)))]`	`mode(v)` chooses the smallest value if two values are tied as most frequent. If you use `[m,f,c]=mode(v)`, then `c{1}` is a vector containing all of the values tied as most frequent.

37. Median of values in vector **v**.

`median(v)`	`median(v)`

38. Basic summary statistics of values in vector **v**.

`summary(v)`	`summary(dataset(v(:)))`

(MATLAB): If you are sure **v** is a column vector as opposed to a row vector, you can simply use `summary(dataset(v))`.

39. Given vector **v**, compute quantiles specified by vector **p** of probabilities.

`quantile(v,p)`	`quantile(v,p)`

Note that the details of how quantiles are computed differ in the two platforms; refer to the documentation for the **quantile** function in each platform for more information. In particular, R allows you to specify a **type** parameter to control how quantiles are computed. Use **type=5** in R to match MATLAB's method of interpolation, i.e., `quantile(v, p, type=5)`.

40. Covariance of paired values in vectors **v** and **w**.

`cov(v,w)`	`tmp=cov(v,w); tmp(2,1)`

(MATLAB): In MATLAB, `cov(v,w)` produces a covariance matrix; extracting the off-diagonal element in row 2, column 1 gives the desired covariance.

41. Covariance matrix for two vectors **v1** and **v2**.

`cov(cbind(v1,v2))`	`cov(v1,v2)`

42. Covariance matrix giving covariances between more than two vectors, e.g., **v1**, **v2**, and **v3**, or between the columns of matrix **A**.

`cov(cbind(v1,v2,v3))` or `cov(A)`	`cov([v1 v2 v3])` or `cov(A)`

(R): You can also use `var(cbind(v1,v2,v3))` or `var(A)`.

(MATLAB): The vectors must be column vectors. If they may be either row vectors or column vectors, use `cov([v1(:) v2(:) v3(:)])`.

43. Given matrices **A** and **B**, build covariance matrix C where the value in row i, column j is the covariance between column i of **A** and column j of **B**.

`cov(A,B)`	MATLAB has a function to do this for correlation coefficients; its results can be used to construct the desired covariances: `[Y,X]=meshgrid(std(B),std(A)); X.*Y.*corr(A,B)`

44. Pearson's linear correlation coefficient between elements of vectors **v** and **w**.

`cor(v,w)`	`corr(v,w)`

(MATLAB): **v** and **w** must be column vectors; use `corr(v(:),w(:))` if they may possibly be row vectors. The same applies for the next two entries.

45. Kendall's tau correlation statistic for vectors **v** and **w**.

`cor(v,w,method='kendall')`	`corr(v,w,'type','kendall')`

46. Spearman's rho correlation statistic for vectors **v** and **w**.

| cor(v,w,method='spearman') | corr(v,w,'type','spearman') |

47. Correlation matrix of pairwise Pearson's correlation coefficient between columns of matrix **A**.

| cor(A) | corr(A) |

The **method** argument may be used in R, and the **type** argument used in MATLAB, as in the previous two items to choose Kendall's tau or Spearman's rho correlation statistics.

48. Given matrices **A** and **B**, build correlation matrix C where the value in row i, column j is Pearon's correlation between column i of **A** and column j of **B**.

| cor(A,B) | corr(A,B) |

The **method** argument may be used in R, and the **type** argument used in MATLAB, as in the previous two items to choose Kendall's tau or Spearman's rho correlation statistics.

4.5 Vectorized logical tests

Section 3.3 (items 26–31) and Section 3.5 (items 74–80) introduced some examples of vectorized logical operators. Additional information is provided here.

Given vectors **v** and **w** containing logical/Boolean/TRUE-FALSE values, both R and MATLAB provide ways to perform pairwise logical operations on the vectors, or operations that apply to all elements within a given vector.

49. AND operation: element **k** of **x** will be TRUE if and only if element **k** of both **v** and **w** are TRUE.

| x = v & w | x = v & w |

50. OR operation: element **k** of **x** will be TRUE if and only if element **k** of either **v** or **w** (or both) are TRUE.

| x = v \| w | x = v \| w |

51. XOR (exclusive-or) operation: element **k** of **x** will be TRUE if and only if element **k** of either **v** or **w** (but not both) are TRUE.

| x = xor(v,w) | x = xor(v,w) |

52. NOT operation: element **k** of **x** will be TRUE if and only if element **k** of **v** is FALSE.

| x = !v | x = ~v |

53. TRUE if all elements of **v** are TRUE, and FALSE otherwise.

| all(v) | all(v) |

See items 57–58 on Page 50 for information about how to apply this function to all rows or columns of a matrix, since the method is the same as when summing matrix entries. In R, you can use **apply** to apply **all** to rows or columns of a matrix. In MATLAB, **all** accepts a parameter indicating the dimension of a matrix to operate along.

54. TRUE if any elements of **v** are TRUE, and FALSE otherwise.

| any(v) | any(v) |

55. In both platforms, when the entries of a vector containing logical values are summed, FALSE counts as 0 and TRUE counts as 1. This lets you easily count the number of entries satisfying certain conditions. For example, the command below counts how many values in **v** are greater than 4 and less than or equal to 7.

| `sum((v > 4) & (v <= 7))` | `sum((v > 4) & (v <= 7))` |

4.6   Other calculations

56. Sum of all elements in vector **v** or matrix **A**.

R	MATLAB
`sum(v) or sum(A)`	`sum(v) or sum(A(:))`

57. Sums of columns of matrix **A**.

| `colSums(A) or apply(A,2,sum)` | `sum(A,1) or just sum(A)` |

58. Sums of rows of matrix **A**.

| `rowSums(A) or apply(A,1,sum)` | `sum(A,2)` |

59. Product of all elements in vector **v** or matrix **A**.

| `prod(v) or prod(A)` | `prod(v) or prod(A(:))` |

60. Products of columns of matrix **A**.

| `apply(A,2,prod)` | `prod(A,1) or just prod(A)` |

61. Products of rows of matrix **A**.

| `apply(A,1,prod)` | `prod(A,2)` |

62. Matrix exponential $e^A = \sum_{k=0}^{\infty} A^k/k!$.

| `expm(Matrix(A))`
This function is in package **Matrix**. | `expm(A)` |

63. Cumulative sum of values in vector **v**.

| `cumsum(v)` | `cumsum(v)` |

64. Cumulative sums down columns of matrix **A**.

| `apply(A, 2, cumsum)` | `cumsum(A,1) or just cumsum(A)` |

65. Cumulative sums across rows of matrix **A**.

| `t(apply(A, 1, cumsum))` | `cumsum(A,2)` |

66. Cumulative sum of all elements in a matrix **A**, going down the columns sequentially.

| `cumsum(A)` | `cumsum(A(:))` |

67. Cumulative product of all elements in a matrix **v**.

cumprod(v)	cumprod(v)

In both platforms, **cumprod** can be used in the same ways as **cumsum** above to compute cumulative products of all elements of a matrix, or the elements down all columns or across all rows.

68. Cumulative minimum or maximum of elements in vector **v**.

cummin(v) or cummax(v)	MATLAB does not currently have a built-in function to do this, but a basic **for** loop can be used. Because MATLAB optimizes/accelerates such loops, this actually runs quite quickly:
	 `w=zeros(size(v)); w(1)=v(1);`
	 `for i=2:length(v)`
	 `  w(i)=min(w(i-1),v(i));`
	 `end`

69. Construct a vector whose i^{th} element is the difference between the $(i + 1)^{st}$ and i^{th} elements of vector **v**, i.e., the difference between consecutive elements of **v**. The resulting vector has one fewer element than **v**.

diff(v)	diff(v)

70. Make a vector **y** the same size as vector **x**, which equals **4** everywhere that **x** is greater than 5, and equals 3 everywhere else.

y = ifelse(x > 5, 4, 3) This is like a vectorized version of the C programming language's ternary operator.	There is no simple built-in function, but it can be accomplished in a couple of different ways via vectorized computations:
	 `z = [3 4];`
	 `y = z((x>5)+1)`
	 Another approach is:
	 `y = 3*ones(size(x)); y(x>5) = 4`

71. Find the minimum element of vector **v**.

min(v)	min(v)

Both platforms also have a corresponding **max** function.

(R): If any of the values in **v** are **NA**, **min(v)** will return **NA**. You can ignore **NA** values by instead using **min(v, na.rm=TRUE)**.

72. Find the minimum element of matrix **A**.

min(A)	min(A(:))

73. Find the minimum elements of each column of matrix **A**.

apply(A,2,min)	min(A)

74. Find the minimum elements of each row of matrix **A**.

apply(A,1,min)	min(A, [ ], 2)

75. Given matrices **A** and **B**, compute a matrix where each element is the minimum of the corresponding elements of **A** and **B**.

pmin(A,B)	min(A,B)

R also has a corresponding **pmax** function.

76. Given matrix **A** and scalar **c**, compute a matrix where each element is the minimum of **c** and the corresponding element of **A**.

pmin(A,c)	min(A,c)

77. Given matrices **A** and **B**, find the minimum value among all elements of both **A** and **B**.

min(A,B)	min([A(:) ; B(:)])

78. Given vector **v**, find the index **ind** of the first element in **v** equal to the minimum element of **v**.

ind = which.min(v)	[y,ind] = min(v)

R also has a corresponding **which.max** function.

79. Given matrix **A**, find the rows **inds** of the minimum elements within each column of **A**.

inds = apply(A, 2, which.min)	[y,inds] = min(A)

80. Given matrix **A**, find the columns **inds** of the minimum elements within each row of **A**.

inds = apply(A, 1, which.min)	[y,inds] = min(A, [ ], 2)

81. Sort elements of the vector **v**.

sort(v)	sort(v)

For complex numbers, R sorts first by the real part, and then by the imaginary part. MATLAB sorts first by the magnitude, and then by the angle.

82. Create sorted vector **s** containing the elements of **v**, with corresponding index vector **idx**, such that s[k] = x[idx[k]].

tmp=sort(v,index.return=TRUE) s=tmp$x; idx=tmp$ix	[s,idx]=sort(v)

83. Given matrix **A**, sort the rows according to the first column. Use column 2 to break ties, then column 3 to break remaining ties, etc. Construct sorted matrix **sA**, and index vector **idx** as in the previous entry, where row **k** of **sA** is row **idx[k]** of **A**.

idx=do.call(order, data.frame(A)) sA=A[idx,]	[sA,idx] = sortrows(A)

84. Sort the rows of **A** according to columns **x**, **y**, and **z**.

idx=order(A[,x],A[,y],A[,z]) sA=A[idx,]	[sA,idx]=sortrows(A,[x y z])

85. Same as previous item, but use decreasing order when sorting by columns **x** and **y**.

`idx=order(-A[,x],-A[,y],A[,z])` `sA=A[idx,]`	`[sA,idx]=sortrows(A[-x -y z])`

86. Given a vector **v** of presumably discrete values, build a vector **w** containing the unique values in **v**, and corresponding vector **c** containing the counts of those values.

`tmp=table(v); c=as.numeric(tmp)` `w=as.numeric(names(tmp))`	`w=unique(v); c=hist(v,w)`

(R): The **table** function actually works with non-numeric data, such as character strings. If you want to build vectors **w** and **c** describing character data, use `w=names(tmp)`, i.e., omit the **as.numeric()**.

87. Given a vector **v** of presumably continuous values, divide the values into k equally sized bins, then build a vector **m** containing the midpoints of the bins and a corresponding vector **c** containing the counts of values in the bins.

`tmp=hist(v,seq(min(v),max(v),`
`    len=k+1),plot=FALSE);`
`m=tmp$mids; c=tmp$counts`
	`[c,m]=hist(v,k)`

88. Compute the convolution of the vectors **x** and **y**.

`convolve(x, rev(y), type='open')`	`conv(x,y)`

5

Lists and Cell Arrays

In both R and MATLAB, a matrix or vector can only hold a set of values of the same type, for example floating-point values, integers, logical values, characters, etc. Sometimes it is convenient to build a more general object which can contain values of different types. Both platforms provide ways of doing this. In R, *lists* are used, while in MATLAB, *cell arrays* are used (but also see MATLAB *structs* in Section 3.8.3). As with matrices and vectors, in R, a list is fundamentally a vector object (from which you can build matrices), while in MATLAB a cell array is fundamentally a matrix (which may be equivalent to a vector if it happens to have only one row or column).

5.1 Creating lists and cell arrays

Let us create a list or cell array containing four elements: (1) a vector **V1** containing the values 4, 8, and 15; (2) a variable **V2** containing the string "hello"; (3) a 2×3 matrix **V3** of random values; and (4) the variable **V4** containing the logical value TRUE. We will build the object in two forms: first, as a vector **tmp**, and second, as a 2×2 matrix **tmpA** (with the four elements arranged down the first column and then the second column).

R

First set up the temporary variables:

```
————————————————————————— R —————————————————————————
V1 = c(4,8,15);  V2 = 'hello'
V3 = matrix(runif(6), nrow=2);   V4 = TRUE
```

As with vectors, the elements of lists can be referred to by number, but if names are provided, then they may also be used to access elements of the list. To create our example list in vector form without names, you can use `tmpNoNames = list(V1, V2, V3, V4)`. If you wish to provide names as well, you can instead use `tmp = list(foo=V1, bar=V2, baz=V3, quux=V4)`. As shown in the example, the names of the list elements do not need to match the variables originally holding their values.

To create the list as a 2×2 matrix, you can either reshape a vector, or use **matrix**. If you have already created the vector list above, you can either do `tmpA = tmp; dim(tmpA) = c(2,2)` or else `tmpA = matrix(tmp, nrow=2)`. Note that both of these methods strip out names of elements that were present in the original list **tmp**. You can restore them via `names(tmpA) = names(tmp)`. In general, the same mechanisms for working with the names of elements of a vector described in Section 3.8.1 can be used with lists.

An alternative way to create a list is to first create an empty list of the desired length and then fill in its elements, as follows.

```
─────────────────────────────────── R ───────────────────────────────────
tmpNoNames = vector('list', 4)
tmpNoNames[[1]] = V1
tmpNoNames[[2]] = V2
tmpNoNames[[3]] = V3
tmpNoNames[[4]] = V4
```

See Section 5.2 for more about the notation involving double square brackets.

MATLAB

First set up the temporary variables:

```
─────────────────────────────── MATLAB ───────────────────────────────
V1 = [4 8 15];  V2 = 'hello'
V3 = rand(2,3);  V4 = true
```

To create a 1×4 cell array (a row vector), use the command tmp = { V1, V2, V3, V4 }. Use semicolons instead of commas between the four elements to create a 4×1 (column vector) cell array.

To create a 2×2 cell array, you can reshape the vector cell array above using tmpA = reshape(tmp, 2, 2). You could also type in the matrix directly, using tmpA = { V1 V3; V2 V4 }. In this case, the values are entered across the rows, rather than down the columns, and so they are typed in a different order than when constructing the **tmp** vector cell array. You could enter the values down the columns by transposing: tmpA = { V1 V2; V3 V4 }'.

Alternatively, you can first create an empty cell array of the desired length, and then fill in its elements, as follows.

```
─────────────────────────────── MATLAB ───────────────────────────────
tmp = cell(1,4);
tmp{1} = V1;
tmp{2} = V2;
tmp{3} = V3;
tmp{4} = V4;
```

See Section 5.2 for more about the notation involving curly braces.

5.2 Using lists and cell arrays

In both platforms, there are two very different methods for extracting the contents of one or more elements of a list or cell array. You can extract elements "bare," i.e., as their fundamental data types. For example, extracting the first element of the example list or cell array in the previous section gives a vector containing the values 4, 8, and 15. You can also extract one or more elements still wrapped within a list or cell array. The danger is that it is easy to make a typo and extract things in the wrong way but not realize it until things start behaving strangely later on. Fortunately, because this issue exists in both platforms, it does not really make it more difficult to transition from one platform to the other.

R

To access elements of a vector list "bare," you can refer to the elements by name or number. You can use double square brackets with names and numbers, or a dollar sign with names only. For the example list **tmp** created in Section 5.1, the following three commands will all access the first element of the list: `tmp$foo`, `tmp[['foo']]`, and `tmp[[1]]`. You can use all three methods to replace an element of the list, e.g., `tmp[[1]] = c(16, 23, 42)`. R also lets you do things like modify part of the vector contained within a list, via commands like `tmp[[1]][2] = 100`, which modifies the second element of the vector stored in the first element of the list.

To access elements of a vector list still within a list, you can use single square brackets, with names or numbers, for example `tmp['foo']` and `tmp[1]`. Notice the slight difference in output when using double and single square brackets. The difference can be made explicit by checking what type of data the two commands return.

```
―――――――――――――――――― R ――――――――――――――――――
> tmp[[1]]
[1]  4  8 15
> tmp[1]
$foo
[1]  4  8 15
> class(tmp[[1]])
[1] "numeric"
> class(tmp[1])
[1] "list"
```

The first command returns a vector of length 3, while the second command returns a list of length 1 whose single element (named **foo**) is a vector of length 3.

MATLAB

To access the elements of a cell array "bare," you use curly braces for matrix indexing. For the example cell array **tmp** from Section 5.1, the command `tmp{1}` will access the first element of the cell array. You can assign a new value to an element of the list in the same way, e.g., `tmp{1} = [16 23 42]`. You can modify part of a vector contained within a cell array, via commands like `tmp{1}(2) = 100`, which modifies the second element of the vector stored in the first element of the cell array.

To access elements of a cell array still contained within a cell array, use the usual parentheses for matrix indexing, for example `tmp(1)`. Notice the difference in output when using curly braces and parentheses, which can be confirmed by checking the types returned by the two expressions:

```
―――――――――――――――― MATLAB ――――――――――――――――
>> tmp{1}
ans =
     4     8     15
>> tmp(1)
ans =
    [1x3 double]
>> class(tmp{1})
ans =
double
>> class(tmp(1))
```

```
ans =
cell
```

The first command returns a 1×3 row vector, while the second command returns a 1×1 cell array whose single element is a 1×3 row vector.

5.3   Applying functions to all elements of lists and cell arrays

Besides simply being useful because they can contain collections of objects of different types or sizes, one of the nice things about lists and cell arrays is that you can also apply functions to all of the data contained within them. Consider two examples. For the first example, say we have a collection **tmp1** which contains three vectors, of lengths 4, 8, and 15, respectively. The three vectors will contain the integers from 11–14, eight equally spaced values between 15 and 18 inclusive, and the squares of the integers from 21–35, respectively. We wish to compute the mean of each of the three vectors. We could of course write a **for** loop to compute the mean of each vector, one at a time. But there is a better way, in both R and MATLAB. For the second example, suppose we have a collection **tmp2**, each element of which is a vector with three elements, **m**, **n**, and **p**. For a given element of **tmp2**, we wish to generate **m** random values chosen from a binomial distribution with parameters **n** and **p**.

R

The functions **lapply**, **sapply**, and **vapply** allow you to apply a specified function to the elements of a list. **lapply** returns its results in a list, while **sapply** attemps to simplify its results to a vector or a matrix if possible. **vapply** lets you specify more information about the type of data produced each time the function is applied to an element of the list, and has a return value similar to **sapply**.

   Example 1: We can set up our sample data via the command `tmp1 = list(11:14, seq(15,18,len=8), (21:35)^2)`. We can then compute the means of the three vectors in **tmp1** in three different ways, as follows.

```
―――――――――――――――――――――――――― R ――――――――――――――――――――――――――
> lapply(tmp1,mean)
[[1]]
[1] 12.5

[[2]]
[1] 16.5

[[3]]
[1] 802.6667

> sapply(tmp1,mean)
[1]   12.5000  16.5000 802.6667
> vapply(tmp1,mean,numeric(1))
[1]   12.5000  16.5000 802.6667
```

lapply returns its results as a list of length 3, where each element contains the desired mean of the corresponding element of **tmp1**. **sapply** returns the same three scalar values, but combined into a vector. (If **lapply** returned a list in which each element contained a vector of the same size, then **sapply** would return a matrix obtained by packing those vectors together as the columns.) **vapply** returns the same thing as **sapply**, but we must explicitly tell it that the function **mean** will return a single numeric value.

Example 2: We can set up an example list via tmp2 = list(c(4, 8, 0.3), c(7, 16, 0.5), c(12, 42, 0.8)). We also need to build a function which takes a vector containing three values, **m**, **n**, and **p**, and returns **m** random values from the binomial distribution with parameters **n** and **p**. This can be done with the following code: f=function(v) {return(rbinom(v[1],v[2],v[3]))}. (See Section 8.1.1 for more information about writing functions.) The **lapply** function then does what we want:

--------------------------------- R ---------------------------------
```
> lapply(tmp2,f)
[[1]]
[1] 2 2 3 4

[[2]]
[1]  9  2  8  5  8  9  8

[[3]]
[1] 36 31 34 33 31 31 33 32 35 35 32 34
```

The results are returned in a list of length 3. Note that using sapply(tmp, f) would give the same results; **sapply** would not be able to simplify the results to a vector or matrix, because the different elements of **tmp2** produced vectors of different lengths.

MATLAB

The function **cellfun** lets you apply a function to the elements of a cell array.

Example 1: We can set up our sample data via the command tmp1 = {11:14, linspace(15, 18, 8), (21:35).^2}. We can then compute the means of the three vectors in **tmp1** as follows.

------------------------------- MATLAB -------------------------------
```
>> cellfun(@mean,tmp1)
ans =
   12.5000   16.5000  802.6667
```

See Section 8.2.7 for information about the **@mean** notation, which builds a function handle to the **mean** function. This provides a way to pass a reference to the **mean** function as an argument to the **cellfun** function.

Example 2: We can set up an example list via tmp2 = { [4 8 0.3], [7 16 0.5], [12 42 0.8] }. Then, we need to write a function which takes a vector [m n p] containing three scalar values **m**, **n**, and **p** and produces **m** random values from a binomial distribution with parameters **n** and **p**. If you have the Statistics Toolbox installed, this will define a function which produces a $1 \times m$ row vector of such values: myfunc = @(v) binornd(v(2), v(3), [1 v(1)]). See Section 8.2.1 for information about this @() notation, which creates what is called an anonymous function that can then be passed to other functions. If you do not have the Statistics Toolbox, you can produce a binomial(n, p) random value by generating n random values from the continuous uniform distribution between 0 and 1 and counting how many of them are less than p. To simultaneously generate m binomial(n, p)

values, create an $n \times m$ matrix of random values and count how many in each column are less than p. This function will do the trick: `myfunc = @(v) sum(rand(v(2),v(1)) < v(3))`. (Note that when you write **sum(A)** in MATLAB with a matrix **A**, it adds the values within each column. In R, **sum(A)** adds up all of the values in the entire matrix.)

There is one additional complication here. The **cellfun** function by default returns its results in a vector or matrix with the same dimensions as the original cell array. However, that only works when the function we provide to **cellfun** returns a scalar value. Here, our function **myfunc** produces vectors containing 4, 7, and 12 values when called with the three elements of our **tmp2** cell array. We can tell **cellfun** to produce a cell array, rather than a regular array or vector, via the **'UniformOutput'** argument, as follows.

```
─────────────────────────── MATLAB ───────────────────────────
>> out=cellfun(myfunc,tmp2,'UniformOutput',false)
out =
    [1x4 double]     [1x7 double]     [1x12 double]
>> out{:}
ans =
      3     2     2     3
ans =
     11     7    10     8     9     9     6
ans =
     33    34    30    37    35    35    32    31    30    32    36    34
```

5.4 Converting other data types to lists and cell arrays

It is necessary sometimes to convert various types of data (such as vectors or matrices) to/from lists and cell arrays. Several common conversion mechanisms of this type are given below.

5.4.1 All values in a numeric vector or matrix

Simple numeric vectors or matrices can be converted to lists or cell arrays quite easily in both platforms.

R

To convert a simple numeric vector **v** to a list, use `myList = as.list(v)`. If the elements of **v** have names, they will be preserved in the list. To convert a matrix **A** to a list, you can use either approach from Section 5.1 for constructing a matrix list: either `myList = as.list(A); dim(myList) = dim(A)` or `myList = matrix(as.list(A), nrow=dim(A)[1])`. Note that names for the elements of **A**, as well as for the rows and columns of **A**, will be lost by either of the above mechanisms. If the elements of **A** were named, you can use `names(myList) = names(A)` to restore them. Use `rownames(myList) = rownames(A)` and `colnames(myList) = colnames(A)` if the original matrix had row and/or column names that you wish to preserve.

MATLAB

Converting a matrix to a cell array in MATLAB is very straightforward, whether it is a vector **v** (a $1 \times n$ or $n \times 1$ matrix) or a full-blown 2-D matrix **A**. Simply use `myCellArray = num2cell(v)` or `myCellArray = num2cell(A)`. If you wish to change the dimensions of the cell array, you can use the **reshape** function.

5.4.2 Matrix, by columns or rows

Suppose you have an $m \times n$ matrix **A**, and wish to split it into either: (1) a list or cell array of length n, where the kth element contains the values from column k of **A**; or (2) a list or cell array of length m, where the kth element contains the values from row k of **A**.

R

If you wish to split **A** into a list whose elements contain the different columns of **A**, use `myList = split(A, col(A))`. If you instead wish to split **A** into a list whose elements contain the rows of **A**, use `myList = split(A, row(A))`.

MATLAB

If you wish to split **A** into a cell array whose elements contain the different columns of **A**, use `myCellArray = num2cell(A,1)`. This will produce a $1 \times n$ cell array, each element of which contains a column vector. If you instead wish to split **A** into cell array containing the different rows of **A**, use `myCellArray = num2cell(A,2)`. This will produce an $m \times 1$ cell array, each element of which contains a row vector.

5.5 Converting lists and cell arrays to other data types

5.5.1 Set of vectors to a single vector

Suppose you have a set **S** (here, "set" will be used to refer to either a list in R or a cell array in MATLAB), where the various elements of the set consist of vectors. The different vectors may or may not be the same length. You wish to combine the sets into a single vector.

R

You can combine all elements of all of the vectors into a single vector via the command v = `unlist(S)`.

MATLAB

The ease with which you can combine these data into a single vector depends on the details of how the vectors within the cell array are arranged, as well as whether the cell array itself is arranged as a row vector or column vector.

In the simplest case, if all elements of the cell array are row vectors and the cell array itself is a row vector (i.e., everything has just 1 row), you can simply use **v = cell2mat(cellArray)**. This will produce a row vector. This also works if the cell array itself, as well as its elements, are column vectors, in which case the result is a column vector.

A bit of trouble arises if your cell array is a row vector but its elements are column vectors, or vice versa. In that case, you must first reshape everything to ensure that they are properly aligned for combining. You can first write a small function which reshapes a given matrix into a row vector. You then apply that function to all elements of your cell array. Next, reshape the cell array itself to ensure that it is a row vector, and then finally convert to a matrix:

```
━━━━━━━━━━━━━━━━ MATLAB ━━━━━━━━━━━━━━━
myfunc = @(A) reshape(A, 1, [ ]);
tmpCellArray = cellfun(myfunc, cellArray, 'UniformOutput', false);
v = cell2mat(reshape(tmpCellArray, 1, [ ]));
```

5.5.2   Set of vectors to matrix

There may be times when you have a set (a list or cell array), where each element of the set is a vector containing k elements, and you wish to turn the data into a matrix. Suppose you wish to turn the ith element of the list into the ith row of the matrix. If you wish to put the data into columns of the matrix, you can simply transpose the result.

R

This situation arises, for example, when using **scan** to read data from a text file (see Section 12.2). You can create a simple example of such a list with two elements, each being a vector containing three values, via L = list(c(4,8,15), c(16,23,42)). Two ways to convert this to a matrix are as follows:

1. do.call(rbind,L). This calls **rbind**, passing it the set of vectors in the elements of **L** as its arguments.

2. t(simplify2array(L)). Note that we must transpose the result of **simplify2array** because it returns a matrix whose i^{th} column contains the vector from the ith element of the list **L**.

The function **simplify2array** is a helper function usually called by **sapply**. The same result could be obtained by going through **sapply**, e.g., by doing sapply(L,c) to just apply the **c** function to each of the vectors contained in the list **L** (which does nothing), and then letting **sapply** put the results together into a matrix.

MATLAB

In MATLAB, unlike R, the orientation (row or column) matters, for both the cell array itself and for the vectors contained in its elements. Assume vectors **v1**, **v2**, and **v3** are all $1 \times n$ row vectors, and that two cell arrays are created via the commands C1 = { v1 v2 v3 }; C2 = { v1 ; v2 ; v3 }. The command cell2mat(C1) will produce a $1 \times 3n$ row vector, equivalent to the matrix produced by the command [v1 v2 v3]. The command cell2mat(C2) will produce a $3 \times n$ matrix, equivalent to the matrix produced by the command [v1 ; v2 ; v3].

Now consider the case where **v1**, **v2**, and **v3** are all $n \times 1$ column vectors instead, and **C1** and **C2** are constructed as above. The command cell2mat(C1) will produce an $n \times 3$ row vector, again equivalent to the matrix produced by the command [v1 v2 v3]. The command cell2mat(C2) will produce a $3n \times 1$ matrix, equivalent to the matrix produced by the command [v1 ; v2 ; v3].

5.5.3 Set of sets to matrix

Suppose we have a set, each element of which is a set of numeric values, and we wish to convert all of this into a matrix. To do this, all of the subsets must have the same length. As in Section 5.5.2, we will place the values from the ith element of the set in the ith row of the matrix.

R

An example of such a list of lists can be created via `L = list(list(4, 8, 15), list(16, 23, 42))`. To convert this to a matrix, we can use the **unlist** function on each element of the list. This would produce a set of vectors. The function **sapply** will let us apply the **unlist** function to every element of our list **L**, and fortunately will gather up the results into a matrix, assuming all elements of **L** are lists of the same length. So we can simply do `t(sapply(L,unlist))`, again needing to transpose because **sapply** puts the resulting vectors into the columns of the matrix, rather than the rows.

MATLAB

Suppose you have a 1-D cell array where each element is a 1-D cell array whose elements are scalars. The **cell2mat** function converts cell arrays to matrices, but it only works when the elements of the cell array are vectors or matrices. So we need to do two conversions.

First, each element of the outer cell array should be converted from a cell array to a vector. This can be done by applying the **cell2mat** function to each element using **cellfun**. Here, because each cell is being converted to a vector, i.e., **cell2mat** is returning a vector rather than a scalar, we must set the **'UniformOutput'** parameter of **cellfun** to **FALSE**, or zero. This command performs this stage of the conversion, assuming **C** is our cell array: `cellfun(@cell2mat,C,'UniformOutput', false)`.

The result of the command above is a cell array whose elements are vectors; one more call to **cell2mat** will convert that to an array. Overall, the command `cell2mat(cellfun( @cell2mat, C, 'UniformOutput', 0))` will therefore convert the cell array of cell arrays to a matrix.

5.5.4 Set of strings to a set of numeric vectors

Suppose you have a set containing strings representing vectors of numeric values. For example, each element of the set may contain one line of text from a file.

R

An example of a list of this type can be created by the command `tmp=list('4 8 15', '16 23 42')`.

The **scan** command can turn a single such string (such as `s='4 8 15'`) into a vector, via the command `scan(text=s)`. However, **scan** will not operate directly on a list of such strings. This is a good candidate for the function **lapply**, which allows us to apply the function **scan** to each element of our list **tmp**. However, **lapply** does not provide a mechanism to specify the name (**text** in our example) of the parameter to pass to the function being called (**scan**). We can get around this by creating a wrapper function which takes a single parameter (the string to scan from), and then in turn calls **scan**, passing that single parameter as the **text** parameter. So this will work: `lapply(tmp, function(x) scan(text=x, quiet=TRUE))`. The **quiet=TRUE** argument to **scan** prevents it from displaying how many records it converted in each line.

MATLAB

An example of such a cell array can be created by the command tmp = {'4 8 15', '16 23 42'}.

The **textscan** function can be used to extract the numerical values from any one of the strings into a new cell array. The **cellfun** function can be used to apply **textscan** to each element of the overall cell array. This is facilitated by first creating a small function which takes a string as a parameter and calls **textscan** with the string as well as the necessary parameter indicating what format to use when scanning the data:

```MATLAB
myfunc = @(s) textscan(s, '%f');
cellfun(myfunc, tmp)
```

6

Flow Control

Both R and MATLAB have similar and standard methods for controlling the flow of code — statements such as **if/else**, **while**, and **for**. The two packages are more similar than different, with mainly just some differences in the syntax used to delimit the loops.

6.1   Conditional ("if") statements

Both platforms have **if** statements to allow a block of code to optionally be executed, depending on the value of a logical test. The common logical operators are listed in Table 6.1.

R

In R, an **if** statement has the form

──────────────────────── R ────────────────────────
```
if (condition) command
```

The **condition** should be a logical (TRUE/FALSE) value; unlike MATLAB, this should be a scalar value. If you provide a vector, only the first element is used, and a warning is displayed. Values other than logicals can be used, as long as they can be interpreted as logicals. For example, non-zero numeric values are interpreted as TRUE while zero is FALSE. The character strings "T," "TRUE," "True," and "true" are interpreted as TRUE, with similar versions for FALSE. The **command** can be on a new line following the **if** statement. Unlike MATLAB, the **condition** must be contained within parentheses. Finally, one additional warning is needed. Although by default the commands T and F will produce the values **TRUE** and **FALSE**, respectively, they can be changed. That is, you can enter

Description	R expression	MATLAB expression
a AND b	a && b	a && b
a OR b	a || b	a || b
a XOR b	xor(a,b)	xor(a,b)
NOT a	!a	~a

TABLE 6.1
Scalar logical operators. The **&&** and **||** operators are short circuiting. That is, when evaluating a logical condition from left to right, **&&** stops evaluating its arguments as soon as any of them are FALSE, and **||** stops evaluating as soon as any of its arguments are TRUE.

the command T = 0, or even T = FALSE. I have seen a graduate student unwittingly set **T** to zero, and then use it later on, thinking its value was still TRUE.[1]

A simple example of an **if** statement is

```
────────────────────────────── R ──────────────────────────────
x = sample(100,1)   # a random integer from 1 to 100
if (x > 50) print('x is large')
```

Because R does not terminate **if** statements with an **end** as MATLAB does, if you wish to put more than one command inside the body of the **if**, you need to enclose the commands within curly brackets:

```
────────────────────────────── R ──────────────────────────────
x = sample(100,1)
if (x > 50) {
  print('x is large')
  x = floor(x/2)
}
```

In MATLAB, you can provide a vector or matrix of logical values as the **condition** in an **if** statement; the commands get executed only if *all* elements of the object are TRUE. In R, you can accomplish the equivalent via the **all** function:

```
────────────────────────────── R ──────────────────────────────
A = matrix(runif(6), nrow=2)   # 2x3 matrix of numeric values
# Note: "A > 0.5" is a matrix of logical values
if (all(A > 0.5))
  print('All values of A were large')
```

MATLAB

In MATLAB, an **if** statement has the form

```
──────────────────────────── MATLAB ────────────────────────────
if condition
  commands
end
```

Unlike R, even if the clause of the **if** statement only contains a single command, you must still use an **end** statement to explicitly flag the end of the **if**. Corresponding MATLAB examples of the **if** statements from Section 6.1 are below. First, a simple example with a single command inside of the **if** statement:

```
──────────────────────────── MATLAB ────────────────────────────
x = randi(100, 1);   % a random integer from 1 to 100
if x > 50
  disp('x is large')
end
```

Multiple statements inside the **if** statement require nothing different:

---

[1] Fortunately, R will not let you do TRUE = FALSE, so it is always safe to compare to **TRUE** and **FALSE** without worry. However, I have heard stories about older versions of FORTRAN in which it was possible to change the values of numerical constants such as 5 via an odd trick.

```
─────────────────────── MATLAB ───────────────────────
x = randi(100, 1);
if x > 50
  disp('x is large')
  x = floor(x/2);
end
```

If the **condition** is a vector or matrix of logical values, the commands get executed only if *all* elements are TRUE:

```
─────────────────────── MATLAB ───────────────────────
A = rand(2,3);   % 2x3 matrix of numeric values
b = (A > 0.5);   % matrix of logical values
if b
  disp('All values of A were large')
end
```

6.2 "If/else" statements

Like most programming languages, both platforms also provide "if/else" statements allowing you to execute one set of statements if a condition evaluates as TRUE, and another set of statements if the condition is FALSE.

R

The syntax for R's **if/else** statement has a possible pitfall that does not arise in MATLAB. If both the TRUE and FALSE sets of code have only one statement, you can write the command like this:

```
─────────────────────── R ───────────────────────
if (condition) command1 else command2
```

If the code for the FALSE clause has multiple statements, you can enclose them in curly braces. And of course even if you have only a single statement (whether in the TRUE or FALSE clauses or both), you can still enclose it in curly braces.

A problem arises in R if you try to put the **else** on a new line, as in the following two code fragments:

```
─────────────────────── R ───────────────────────
if (condition)
  command1
else
  command2


if (condition) {
  command1a
  command1b
}
else {
```

```
    command2a
    command2b
}
```

The problem is that when R is parsing its input line by line, it thinks the **if** statement is done before it gets to the **else**. So the safe way to write an **if/else** statement in R is as follows, which keeps the **else** on the same line as the curly brace which terminates the TRUE clause of the statement.

```
────────────────────────────── R ──────────────────────────────
if (condition) {
  commands1
} else {
  commands2
}
```

For safety and simplicity, some people recommend the use of curly braces even when the **if** or **else** contains only one command, following the pattern in the block of code above.

Testing against multiple cases is written as follows:

```
────────────────────────────── R ──────────────────────────────
if (x < 5) {
  print('x is very small')
} else if (x < 10) {
  print('x is somewhat small')
} else if (x < 50) {
  print('x is medium')
} else {
  print('x is large')
}
```

Leaving out any of the sets of curly braces except for the last pair would cause an error; the pair of braces around the final clause is not strictly necessary in the above example.

MATLAB

MATLAB's **if/else** statements do not have the potential problem that R's do, because MATLAB always forces you to terminate an **if** statement with an **end**, no matter how many commands are inside the TRUE/FALSE clauses. So the code looks like this:

```
──────────────────────────── MATLAB ────────────────────────────
if (condition)
  commands1
else
  commands2
end
```

MATLAB also provides the **elseif** mechanism for testing against multiple cases. For example:

```
──────────────────────────── MATLAB ────────────────────────────
if x < 5
  disp('x is very small')
elseif x < 10
  disp('x is somewhat small')
```

```
elseif x < 50
  disp('x is medium')
else
  disp('x is large')
end
```

6.3 "for" loops

Both platforms have **for** loops which behave in a very similar manner. They differ somewhat from how **for** loops work in languages like C, C++, and Java. R and MATLAB's **for** loops are more similar to **foreach** loops in Perl, in that you specify a set of values for the dummy/index variable to take on, and changing that variable's value inside of the loop does not affect how many iterations are performed.

R

The syntax for a loop in R is:

———————————————— R ————————————————
```
for (ind in vec)  command
```

where **ind** is the name of a dummy variable (you can use any legitimate variable name here), and **vec** is a vector of values. This is equivalent to the set of statements

———————————————— R ————————————————
```
ind = vec[1]
command
ind = vec[2]
command
...
ind = vec[N]
command
```

where **N** is the length of the vector **vec**. If multiple statements are to be included inside the loop, enclose them with curly braces. For example, to add the values from 1 to 100, as well as their squares:

———————————————— R ————————————————
```
mySum = 0
mySumSqr = 0
for (i in 1:100) {
  mySum = mySum + i
  mySumSqr = mySumSqr + i^2
}
print(sprintf('sum of values is %d', mySum))
print(sprintf('sum of squaresis %d', mySumSqr))
```

One thing to watch out for in R is that some commands which would normally print output when typed at the command prompt will be silent if you put them inside of a **for** loop. For example, if you insert the simple statement **mySum** inside the **for** loop in the above code to see how the value of **mySum** changes over time, you will not see any output. Instead you should use `print(mySum)`. The same is true for commands executed from script files (see Section 7.3).

If you pass a matrix **A** in as the set of values a **for** loop, using a command like **for (i in A)**, it will sequentially loop over each individual element of **A**, one column at a time. That is, it is equivalent to the command **for (i in c(A))**.

MATLAB

As with **if** statements, MATLAB terminates the body of a loop with an explicit **end** statement, even if the loop only contains one command. The syntax for a loop in MATLAB is:

```
──────────────────── MATLAB ────────────────────
for ind=vec
   command
end
```

where **ind** is the name of a dummy variable (you can use any legitimate variable name here), and **vec** is a vector of values. This is equivalent to the set of statements

```
──────────────────── MATLAB ────────────────────
ind = vec(1)
command
ind = vec(2)
command
...
ind = vec(N)
command
```

where **N** is the length of the vector **vec**. For example, to add the values from 1 to 100 as well as their squares:

```
──────────────────── MATLAB ────────────────────
mySum = 0;
mySumSqr = 0;
for i = 1:100
  mySum = mySum + i;
  mySumSqr = mySumSqr + i^2;
end
disp(sprintf('sum of values is %d', mySum))
disp(sprintf('sum of squaresis %d', mySumSqr))
```

One important thing to watch out for with MATLAB **for** loops is that the vector of values you provide must be a row vector, rather than a column vector. That is, **for i=[1 2 3 4 5]** is a loop which will have five iterations (with **i** being a scalar value inside each iteration), while **for i=[1; 2; 3; 4; 5]** will have only one iteration (with **i** being a 5×1 vector in that one iteration). Fortunately, the most common way of constructing the values for **for** loops, namely expressions like **a:b** or **linspace(a, b, n)** return row vectors.

More generally, if you write **for i=A**, where **A** is an array, the loop will iterate with **i** being assigned sequential columns of **A** throughout the loop, as shown below:

```
──────────────────── MATLAB ────────────────────
>> A=[4 8 15 ; 16 23 42]
A =
```

```
     4      8     15
    16     23     42
>> mySum=0;
>> for i=A
  i
  mySum = mySum+i;
end
i =
      4
     16
i =
      8
     23
i =
     15
     42
>> mySum
mySum =
     27
     81
```

6.4 "while" loops

Both platforms have **while** loops, which allow you to execute blocks of code as long as a given condition is satisfied. A very simple example would be to repeatedly choose random values between 0 and 1 until a value larger than 0.95 is chosen, and display how many tries it took until such a value was found.

R

The R syntax for a **while** loop is **while (condition) expression**. The condition must be enclosed in parentheses. If the expression consists of more than one command, it should be enclosed in curly braces. Here is code to implement the example:

R

```
numTries = 1
while (runif(1) <= 0.95) {
  numTries = numTries + 1
}
print(numTries)
```

Note that I have used curly braces here, even though there is only one statement inside the **while** loop. I tend to use curly braces most of the time, to guard against accidentally leaving them out if I add additional commands to a loop which previously contained only one command.

MATLAB

Like **if** statements and **for** loops, MATLAB's **while** statements must be terminated with an **end** to delimit the commands enclosed within the loop. Here is code for the example:

```MATLAB
numTries = 1;
while rand < 0.95
  numTries = numTries + 1;
end
numTries
```

6.5   Breaking out of loops

Both platforms provide a way to either entirely stop executing a **for** or **while** loop early, or to stop executing just the current iteration and continue with the next iteration. A **break** statement inside of a **for** loop will exit the entire loop immediately in both R and MATLAB. If you have nested loops, the **break** exits only the innermost loop.

To stop execution of just the current iteration of the loop and begin the next iteration, in R you can use the command **next**, while the MATLAB equivalent is **continue**. Sample code is given below which simulates the repeated flipping of two coins to generate one of the values 1, 2, or 3. Any of the outcomes tails/tails, tails/heads, or heads/tails increments one of three counters; the outcome heads/heads is ignored. The process is repeated until one of the three counters reaches 10, in which case the "winning" counter is displayed. However, if the sum of the three counters ever exceeds 20, the loop is terminated early.

R

```R
# Flip two coins to choose a value from 1..3.  Repeat until one of
# the values has been seen 10 times.
counts = rep(0, 3)
while (max(counts) < 10) {
  if (sum(counts) > 20) {
    print('Stopping early!')
    break
  }
  coin1 = sample(c(0,1), 1)  # 0=tails, 1=heads
  coin2 = sample(c(0,1), 1)
  if (coin1 + coin2 == 2) # skip if both are heads
    next
  index = coin1 + 2*coin2 + 1  # index of 0, 1, or 2 for T/T, H/T, T/H
  counts[index] = counts[index] + 1
}
print(counts)
if (max(counts) == 10) {  # display winner if anyone won
  winner = which(counts == 10)
  print(sprintf('Counter # %d won', winner))
}
```

Note that unlike MATLAB, R also has a **repeat** command, equivalent to **while (TRUE)**. It is mentioned here, because the only way to escape from a **repeat** loop is via **break**. For example, here is how to write the example from Section 6.4 which counts how many random values are generated until one larger than 0.95 is found:

—————————————————————— R ——————————————————————

```r
numTries = 1
repeat {
  if (runif(1) > 0.95)
    break
  numTries = numTries + 1
}
print(numTries)
```

MATLAB
—————————————————— MATLAB ——————————————————

```matlab
% Flip two coins to choose a value from 1..3.  Repeat until one of
% the values has been seen 10 times.
counts = zeros(1,3);
while max(counts) < 10
  if sum(counts) > 20
    disp('Stopping early!')
    break
  end
  coin1 = randi([0 1]);  % 0=tails, 1=heads
  coin2 = randi([0 1]);
  if coin1 + coin2 == 2  % skip if both are heads
    continue
  end
  index = coin1 + 2*coin2 + 1; % index of 0, 1, or 2 for T/T, H/T, T/H
  counts(index) = counts(index) + 1;
end
counts
if (max(counts) == 10)  % display winner if anyone won
  winner = find(counts == 10);
  disp(sprintf('Counter # %d won', winner))
end
```

6.6 "switch" statements

In some programming languages such as C, C++, and Java, a **switch** statement provides a convenient way to excute one set of commands among several sets, according to the value of a variable. It is equivalent to a set of **if ... else** statements, but more concise.

R

R's **switch** statement is actually quite different from MATLAB's and the one in C/C++/Java. In R, **switch** is used to select one of a set of values, rather than to perform a set of commands. However, it can be fairly easily coerced into behaving in a manner somewhat similar to MATLAB's.

R's **switch** statement behaves differently according to whether you provide it with a character string or a numeric value as its first argument. If the first argument is not a string, the value gets coerced to an integer (floating point values, for example, get truncated to integers). The return value of the command **switch(x, a, b, c, ...)** is one of the values **a, b, c, ...** according to whether **x** is 1, 2, 3, ... If you provide k arguments in addition to the parameter **x**, but the value of **x** is greater than k, then NULL is returned. Here is an example that lets you set **y** equal to one of the first five prime numbers, according to whether **x** is 1, 2, 3, 4, or 5:

```
——————————————————— R ———————————————————
> x = 4
> y = switch(x, 2, 3, 5, 7, 11)
> y
[1] 7
> x = 6
> y = switch(x, 2, 3, 5, 7, 11)
> y
NULL
```

Note that the results of the above **switch** commands are almost identical to what would be achieved via the command v = c(2,3,5,7,11); y = v[x]. However, where **switch** returns **NULL** if no match is made, the command **v[x]** would give **NA** instead.

switch is capable of some other behaviors as well. If the first parameter to **switch** is a character string, it is matched against the names of the other parameters. Only exact matches count. For example:

```
——————————————————— R ———————————————————
> x='beta'
> switch(x,alpha=101,beta=102,gamma=103)
[1] 102
> x='delta'
> y=switch(x,alpha=101,beta=102,gamma=103)
> y  # note, if you just call switch directly, you won't see the NULL
NULL
```

If there is more than one match, the first one is used. If one of the arguments is unnamed, it will be used as the value of the command if the first parameter does not match any of the other parameter names:

```
——————————————————— R ———————————————————
> x='beta'
> switch(x,alpha=11,99,beta=12,gamma=13)
[1] 12
> x='delta'
> switch(x,alpha=11,99,beta=12,gamma=13)
[1] 99
```

If the value of the matching parameter is missing, the next non-missing parameter's value is used:

```R
──────────────────────── R ────────────────────────
> x='alpha'
> switch(x,alpha=,beta=,gamma=17,99)
[1] 17
```

As mentioned above, R's **switch** statement can be coerced into executing one of a
different set of commands according to the value of a variable. This is achieved by making
the various parameters be blocks of code enclosed in curly braces:

```R
──────────────────────── R ────────────────────────
> x=0; z=0
> y = 'beta'
> switch(y, alpha={x=12; z=17},
+         beta={x=57; z=42},
+         gamma={x=17; z=121})
> x
[1] 57
> z
[1] 42
```

If you use a numeric variable as the first argument, then if that variable evaluates to the
value **i**, the ith block of code will be executed.

MATLAB

MATLAB's **switch** statement is more similar to that of C/C++/Java. One of a set of
commands will be executed according to which value an expression matches. For example:

```MATLAB
──────────────────────── MATLAB ────────────────────────
x = 4
switch x
  case 1
    disp('x is small')
    y = -1;
  case {2, 3, 4}
    disp('x is medium')
    y = 0;
  case 5
    disp('x is large')
    y = 1;
  otherwise
    disp('x outside range 1..5')
    y = -99;
end
```

When the above code is executed, it displays "x is medium" and sets **y** to zero. The above
code demonstrates that to match against a set of multiple values, those values should be
combined in a cell array. It also shows that you can have an **otherwise** clause (equivalent
to C's **default** case). Finally, also note that you do not need **break** statements at the end
of each block of commands, as you do in the C programming language.

String values can also be used in **switch** statements.

6.7   "ifelse" statements in R

R has an additional statement called **ifelse**, which provides a way to select values from two vectors. You provide a vector **test**, along with vectors **yes** and **no**. The return value is a vector the same size as **test**, where the ith element of the vector is chosen from **yes** if the ith element of **test** is TRUE, and is otherwise chosen from **no**. For example:

```R
> v = c(TRUE, FALSE, TRUE, TRUE, FALSE)
> first = 11:15
> second = 71:75
> ifelse(v, first, second)
[1] 11 72 13 14 75
```

The same trick could be achieved by turning the TRUE/FALSE values into 1 and 2, respectively (this can be done using the expression **2-v**), and then using matrix indexing of a matrix following entry 64 from Section 3.5:

```R
> cbind(first, second)[cbind(1:length(v), 2-v)]
[1] 11 72 13 14 75
```

However, this latter method is admittedly harder to make sense of.

Although MATLAB does not have a direct equivalent of R's **ifelse** statement, the same result can be achieved with matrix indexing similar to the alternative approach for R given above. First, a temporary matrix can be built, whose columns contain the vectors **first** and **second**. Then, column indices to choose elements from either the first or second column can be constructed from **v**:

```MATLAB
>> v = [true,false,true,true,false];
>> first=11:15;
>> second=71:75;
>> tmp=[first(:) second(:)]
tmp =
     11    71
     12    72
     13    73
     14    74
     15    75
>> tmp(sub2ind(size(tmp), 1:length(v), 2-v))
ans =
     11    72    13    14    75
```

7

Running Code from Files: Scripts

If you will be entering more than a small handful of commands, rather than simply typing them at the command prompt, it is prudent to use an editor to place them in a file. This allows you to fix any errors in your code without needing to retype all of your commands. It is also very useful if you are experimenting with some properties of a figure to get it just right; when you have it just the way you like, you can save your commands to a file so that you can easily reproduce or modify the figure again later.

Both R and MATLAB have built-in editors[1] and ways to execute commands from their built-in editors and from files. Following standard computer terminology, a file (or editor window) containing some R or MATLAB commands is typically called a *script*.

7.1 Current working directory

Both R and MATLAB use what is called a *current working directory*.[2] The current working directory is the default location for reading and writing files.

R

The command `getwd()` will display the current working directory. In the OS-X version of R, it is also displayed at the top of the Console window. You can use `setwd('some/path/here')` to set the current working directory to the specified path. Use forward slashes as the path separators. Tilde expansion will be done on systems that support it, where a tilde represents your home directory. (You can enter `path.expand('~')` to see what R considers your home directory to be.) So for example, on most platforms you can enter a command like `setwd('~/R-documents')`, and under an appropriate version of Windows you could enter a command like `setwd('C:/Users/hiebeler/Documents/R-documents')`.

You can see a listing of the files in the current working directory via the command `dir()`, and can make a new directory (folder) named **foo** via `dir.create('foo')`.

MATLAB

The command `pwd` shows (prints) the current working directory, or you can also use `cd` with no arguments. You can use **cd** with an argument to set the working directory, e.g., `cd some/path/here`. Tilde expansion will be done on some operating systems. Note that if your path contains directories with spaces in them, you'll need to use the function-call form of

---

[1] The standard Linux version of R does not have a GUI with an editor, but there are facilities to integrate the Emacs editor with R. Also see the free and open source RStudio project.

[2] "Directory" is the traditional Unix name for what many people refer to as a "folder" under Windows and Mac OS-X.

cd, for example cd('C:/Users/David Hiebeler/Desktop'). And note that cd(dir) sets the current working directory to the path stored in the string variable named **dir**, and not to a directory named **dir**.

You can see a listing of the files in the current working directory via the command dir, and can make a new directory (folder) named **foo** via mkdir foo.

7.2   The MATLAB search path

In addition to the current working directory, MATLAB also has a *search path*, which is a set of directories it searches when looking for commands. This can be useful if you create some scripts or functions that you may want to use regularly. Entering the command path will show you the current search path; this may be quite long if you have many toolboxes installed. The **addpath** command can be used to add a new folder to the search path. For example, addpath ~/myMatlabStuff -end will add the specified folder to the end of the search path (append **-begin** to the command to place the folder at the beginning of the search path instead). You could remove that folder from your path via the command rmpath ~/myMatlabStuff. You can also use the GUI to edit the search path; simply click on the "Set path" menu item (this is under the File menu in older versions of MATLAB). After modifying the path, you can either click "Save" in the GUI window or use the **savepath** command to save your changes so that they will still be used the next time you run MAT-LAB.

Sometimes you may not be sure where in your path a file resides. For example, you may be able to call a function **foo1**, but do not know where the file is. You can use the command which foo1 to find the location of a file on your path. This is especially useful if you accidentally gave a variable the same name as a function, such as **sqrt**. If you enter the command sqrt=8, then the command sqrt(2) will no longer compute $\sqrt{2}$. (In fact, it will give an error, thinking you are trying to access the second element of your variable **sqrt**, which only has one element.) The command which sqrt will tell you that **sqrt** is a variable. If you clear sqrt, then which sqrt will tell you that it is a built-in function again. You can also use the command edit foo1 to open up the M-file defining **foo1** in MATLAB's built-in editor (or another editor, if you have set one in the preferences), if it can be found in the search path. The command type foo1 will display the contents of the M-file right in the main MATLAB window.

Note that R does not have a search path that it uses when executing code from files.

7.3   Executing code from a file

For now, let us assume that you have already created a text file containing some R or MATLAB commands. How you execute the contents of the file is quite different between the two platforms.

R

In R, executing the contents of a file is called "sourcing" the file. You can do this by selecting "File → Source File" (Mac) from the user-interface menus or "File → Source R code" (Windows).[3] A file browser window will pop up, and you can choose the file whose contents you wish to execute.

You can also source the contents of a file by using the **source** function at the command prompt, for example, `source('dostuff.R')`. That command will look for the file in the current working directory (see Section 7.1). You can also specify a partial or full pathname of a file when using the **source** command, such as `source('myFiles/dostuff2.R')` or `source('~/dostuff3.R')`.

In most ways, sourcing a file is equivalent to typing (or copying and pasting) the contents of the file at the R command prompt. The main exception is that some commands which produce output when entered at the R console will be silent when they are executed by sourcing a file. For example, entering `y=7*8; y` at the command prompt will show the result, 56. If you save those commands in a file and **source** it, you will not see any output. There are a couple of ways to address this issue. First, you can add the **print.eval=TRUE** argument to the command, i.e., use `source('dostuff.R', print.eval=TRUE)`. Alternatively, you can use `source('dostuff.R', echo=TRUE)` to see all of the commands themselves in your file, as well as their output, as they are executed. The other way to make the output of commands visible when sourcing a file is to use the **print** function on any expressions whose output you wish to see. For the example commands above, you would change them to `y=7*8; print(y)`. Then you would see the value of **y** whether you enter the commands in the R console, or source them from a file. This same issue arises when you put commands inside of a function, or a loop such as a **for** or **while** loop. Also note that this issue applies to some graphical output, including calls to lattice graphics functions such as **levelplot**. You should use **print(levelplot(...))** if you are executing the commands by sourcing a file. Forgetting to **print** inside a script or a loop is a common source of frustration when your code is not producing the expected output!

MATLAB

Executing commands from a file in MATLAB is much simpler than in R; you simply need to enter the name of the file containing the commands (without the ".m" suffix). For example, if there is a file **dostuff.m** in the current working directory or in the search path (see Sections 7.1 and 7.2), then entering the command `dostuff` at the MATLAB command prompt will execute the contents of the file, just as if you had typed them in directly.

7.4   Creating a new script document in the editor

R

You can use the "File → New Document" (Mac) or "File → New Script" (Windows) menu item to open up a new window for editing code.

---

[3]Note that in Windows, this menu item is only visible in the File menu if the R Console window is active. That is, you should click on that window first to be sure it has the mouse's focus, as opposed to an editor window or a figure window.

MATLAB

Under newer versions of MATLAB, you can use the "New → Script" menu item to open up a new window for editing code; in older versions of MATLAB, use the "File → New → Blank M-File" menu item.

7.5   Comments in script files

R

In R, the comment character is #. Anything from that character to the end of the line will be ignored. So you may have a short script that looks like this:

```
───────────────────────────── R ──────────────────────────────
# set up a variable with Avogadro's constant

avogadro = 6.02e23    # three significant digits, in scientific notation
```

You can type comments directly at the R command prompt, but there is usually not much reason to; comments are primarily used in scripts and functions.

MATLAB

In MATLAB, the comment character is %. Anything from that character to the end of the line will be ignored. So you may have a short script that looks like this:

```
─────────────────────────── MATLAB ────────────────────────────
% set up a variable with Avogadro's constant

% Here is the actual variable definition:
avogadro = 6.02e23   % three significant digits, in scientific notation
```

As in R, you can type comments directly at the MATLAB command prompt, but comments are primarily used in scripts and functions.

MATLAB also has another convention for comments in scripts. If the above script is stored in the file **avo.m** and is in your search path, then typing `help avo` at the command prompt will display all comments up to the first line which does not begin with a comment (whether that is a blank line, or a line containing some commands). The percent signs are stripped out when displaying the comments. For the example script above, you would see the following:

```
─────────────────────────── MATLAB ────────────────────────────
>> help avo
   set up a variable with Avogadro's constant
```

7.6 Executing code from the editor window

R

You can execute selected code from your R editor window. Note that you must first ensure that the editor window containing the code you want to execute is the active window, i.e., you should click on that window to make sure it has the application's focus.

You can select some text in your editor window (perhaps all of it), and then choose "Edit → Execute" (Mac) or "Edit → Run line or selection" (Windows). R will essentially copy and paste the selected text into the command window, as if you had typed it yourself at the prompt.[4] In Windows, using "Edit → run line or selection" when you have no text selected will simply execute the line of code containing the cursor; also, you can use "Edit → Run all" to execute the entire contents of the editor window.

Because executing code from an editor window as above essentially copies and pastes the selected code into the R console, there is one danger to be aware of. If your code uses **scan** to read input from the user, it will not work correctly. Consider the following short code:

──────────────────────── R ────────────────────────
```
x = scan()
y = 17
```

If you execute this code from an editor window, rather than reading the input for **x** from the keyboard, it will try to read the text "**y = 17**". This will fail because it is trying to read numeric data. This may come as a surprise, since MATLAB does not exhibit this behavior with its **input** statement. Note that this problem with **scan()** does not occur if you use the **source** command to execute your code from the file.

Under Mac OS-X, you can also use "Edit → Source Document" to execute the entire contents of the editor window by sourcing it, as described in Section 7.3. If your file has not been changed since you last saved it, R will simply **source** your file. If the file has not been saved since you last modified it in the editor, R will create a temporary file which contains the contents of your editor window, source that file, and then "unlink" (remove) the temporary file. Keep in mind that the same issues that arise when you **source** a file will occur here, namely you will not see the output of various commands whose results you would see if you typed them directly at the command prompt. Also, if R needs to create and then unlink a temporary file to source, you will have the problem with **scan** mentioned above.

MATLAB

To run code within the editor without saving it, you can first delineate a block or section of code by creating lines in the document that each only contains the special comment string %%. All lines between any pair of such lines will be a code block. You can execute a code block by moving the cursor within that block, and then from the Editor toolbar at the top of the editor window selecting "Run Section." If the document has only a single line which contains just the string %%, all lines above this special comment line will be one section, while all lines below will be another section. In slightly older versions of MATLAB, you could first select some or all of the text, and then from the Editor toolbar select "Run

─────────────

[4]One slight difference is that on Mac OS-X, after executing code in this way, if you then press the Up-Arrow key at the command prompt, it will insert all of the code that you executed, rather than just the last line. You can then edit the block of commands before pressing <Enter> to execute them.

Section." In even older versions of MATLAB, the menu item was called "Text → Evaluate Selection."

To save the code and run it, you can select "Run" from the Editor toolbar, or in older versions, use "Debug → Save File and Run" (if you have already saved the file as **dostuff.m**, this menu item will change to "Debug → Run dostuff"). Note that if you save the file and run it as described above, if you save the file in a folder other than the current working directory and that folder is not in the search path, MATLAB will give you the option to either add that folder to the search path or make that folder the current working directory.

7.7 Summary of differences

- In MATLAB, you simply type the name of a script file (without the ".m" suffix) to execute its contents. In R, you must use a **source** command to read and execute the contents of a file. This makes MATLAB feel more customizable, as running commands from a file is syntactically identical with running a built-in MATLAB command with no arguments. By setting up your search path in MATLAB, you can keep a folder containing often-used or useful scripts, which MATLAB will automatically find.

- In R, commands which display their output when typed at the command prompt are silent when executed from within a script. Calls to lattice graphics functions also produce no output when called from within a script. You must use **print** (or another workaround) to make such output visible.

8

Writing Your Own Functions

Putting commands in a script is a useful way to build up more complex code, with an easy way to modify and debug your commands. But sometimes you want a convenient way to modify the behavior of your script by specifying some parameters. For this, the easiest way is to write a *function*, which takes zero or more parameters as inputs, and returns zero or more values as outputs. A function with no formal returned outputs can still be very useful, because it may produce what are called *side effects*. For example, it may produce a plot, or write some data to a file.

8.1   R

8.1.1   Writing functions

In R, a function is technically stored in a variable, whose name corresponds to the name of the function. For example, at the command prompt you can type: `f=function(x) x*x`. If you then type `f`, you will see the function definition. You can call it by entering a command such as `f(11)`, which will return the value 121. You can use your function in expressions, just as any of R's built-in functions, e.g., `y=sin(f(11))`.

You do not need to put the definition of your function `f` into a file called **f.R** (analogous to what is typically done in MATLAB). But for functions which are more than a few lines long, it is generally recommended to put the code in a file, for easier editing and debugging. For example, consider the file **fib.R** below, which defines a function which computes Fibonacci numbers defined by $F_{n+1} = F_n + F_{n-1}$. Note that the standard Fibonacci sequence begins with $F_0 = 0$ and $F_1 = 1$.

```
                                    fib.R
1   # Compute Fibonacci numbers
2   #  steps: how many iterations to do -- if steps=k then compute the
3   #         (k+1)'st Fibonacci number
4   #  first,second: the first two numbers in the sequence
5   fib=function(steps,first=0,second=1) {
6     for (i in 1:steps) {
7       nextval = first+second  # compute next value
8       first = second  # update records
9       second = nextval
10    }
11    return(second)
12  }
```

Some things to note about functions in R are listed below.

1. Default values can be provided for some or all parameters; this is done for the

last two of the three parameters in the **fib** function above. If values for **first** and **second** are not provided when the function is called, those default values (0 and 1, respectively) will be used. You can check whether a value was explicitly provided for a parameter by using the function **missing**. For example, in the **fib** function, you could use if (missing(second)) to test whether a value for the **second** parameter was given. This can be useful for altering how one parameter is used based on whether another parameter is given (e.g., a plotting function that can receive both **x** and **y** data, but behaves differently if only **y** data are provided).

2. Unlike MATLAB, no information about return values of the function is provided at the top of the definition.

3. A **return** statement is used to return a value from the function. This is not strictly necessary; if the end of your function is reached without encountering a **return** statement, then the result of the last evaluated expression will be returned. For example, line 11 in the above function could be changed to simply read second. Personally, I think it is best to always explicitly call **return**, as it highlights all of the locations where execution may exit your function, and explicitly indicates just what value you want returned to the caller.

4. Expressions which would normally display output when typed at the R command prompt will instead be silent when executed inside of a function. This behavior is the same as for R code executed from a script (see Page 79). If you want to see the contents of the variable **myVar** from within a function, you need to use print(myVar) rather than simply using the expression myVar. And in particular, if you use lattice graphics, you need to embed those commands in **print** calls as well in order for the graphics to be displayed. I use **levelplot** from within functions quite often, but I still frequently forget to do this.

5. You cannot return multiple values to be stored in separate variables by the caller, as in MATLAB where you can do [a,b]=f(x). If you wish to do something similar, the easiest thing to do is return a list (see Section 5.1) containing the various items via a command like return(list(val1,val2)), and then let the caller extract the elements of the list and store them in the different variables.

6. Assuming the code for your function is defined in a file, such as **fib.R** in this example, if you modify the code, you must remember to first save the file, and then explicitly source('fib.R') to read in the updated definition. It does not happen automatically, as it does in MATLAB. If you are a long-time MATLAB user, it can take a while to train yourself to remember to do this and cause quite a bit of frustration when you modify your function but do not see the new behavior when calling it.

7. You can write a function (which may have some side effects, such as creating a plot) which does not print anything if its return value is not assigned. This is done via the function **invisible**. For example, if line 11 of **fib.R** is changed to return(invisible(second)), then entering x=fib(15) stores the value 987 in **x**, but fib(15) produces no visible output.[1]

---

[1] Calling fib(15) and entering .Last.value will show that the value 987 was in fact invisibly returned. Or, you can use the parentheses trick from Page 10 and use (fib(15)) to see the invisible return value, or simply print(fib(15)).

8.1.2 Calling functions

When calling functions in R, values of parameters may be given either by order/position or by name. For an example of giving parameters by order, one may enter `fib(10,2,1)` to compute the tenth Lucas number[2] after the first two, namely $L_{11} = 199$. The three parameters in the call to **fib** specify the values of the three arguments **steps**, **first**, and **second** that the function **fib** accepts, in that order.

You can instead specify parameters by name, in any order. For example, calling the function as `fib(second=1, steps=10, first=2)` will also compute L_{11}, as above. You can abbreviate parameter names, as long as the abbreviation is unique; for example, `fib(se=1,st=10,f=2)`. If the abbreviation is not unique, as in `fib(s=5)`, R will give an error. The full rules are a bit more complicated than that — exact matches take a parameter out of consideration for partial matching. For example, `fib(steps=10,s=5)` will work, because **steps** has an exact match, so the only thing **s** can match is the parameter named **second**. A unique partial match does not give the same behavior, i.e., `fib(st=10,s=5)` will fail.

You can mix and match order- and name-based parameters. For example, the command `fib(first=2,10,1)` gives the named parameter value **first** the value 2. There are then two remaining unnamed parameter values, 10 and 1. They are assigned to the remaining unassigned arguments, namely **steps** and **second**, in that order. I personally try to avoid mixing the two methods of passing parameters in this way, with the following exception: I may specify some leading parameters by order (i.e., the first through the nth parameters), and then specify additional parameters by name. For example, `fib(20, second=4)` to specify the first parameter (**steps**) and the value for **second**, while accepting the default value for **first**.

Note that passing parameters by name to functions is where the difference between the assignment operators `=` and `<-` becomes very important in R. Consider the following examples: `f(x=12)` calls the function **f**, specifying that its argument **x** should have the value 12. The command `f(x<-12)` first assigns the value 12 to the variable **x**, and then also uses the value 12 as the first argument to the function **f**, whatever that parameter may be called in **f**. So the very misleading `fib(second<-4)` would assign the value 4 to the variable named **second**, and then call **fib** with a value of 4 as the first argument (namely **steps**), and the default values of the other two parameters (**first**=0 and **second**=1). Be aware that this type of command, where one of the arguments has a side effect (such as assigning a value to a variable in the above examples) may give surprising results, depending on the function **f**. R has a "lazy" method of evaluating function arguments, only doing so when needed [26]. For example, if you have defined `f = function(a,b) return(42+b)`, then `f(x<-12, y<-10)` will assign the value 10 to **y**, but will not assign the value 12 to **x**. This is because the first argument is not actually used within the function, and therefore the expression "**x<-12**" does not get evaluated.

If you have used MATLAB for some time, one final difference you may notice with R is that you must provide an empty parameter list (empty parentheses) when calling a function which takes no parameters. For example, you must use the command `date()`, rather than just `date`.

8.1.3 Environments and variable scope

As in many languages, variables defined within a function are only visible within that function, and vanish when the function exits. For example, if you define the function

---

[2]Lucas numbers are a Fibonacci sequence defined via $L_{n+1} = L_n + L_{n-1}$, with $L_0 = 2$ and $L_1 = 1$.

```
─────────────────────────────── R ───────────────────────────────
f1 = function(x) {
  y = 12
  return(y*x)
}
```

then the variable **y** has not been set at the top-level or global scope. If you enter `f1(8)`, then subsequently entering y will not show the value 12. When the function **f1** exits, the variable **y** created inside of it disappears. If you already had a variable **y** defined before calling the function, its contents will not be modified. This behavior is common to both R and MATLAB (and many other programming languages as well).

There are three types of variables that can exist within a function. The first type is a *formal parameter*, which is a variable defined as an input to the function. In our earlier **fib** function, the formal parameters are **steps**, **first**, and **second**. The second type is a *local variable*, i.e., a variable defined inside of the function. In **fib**, we had a local variable **i** used in the **for** loop, while the function **f1** above has a local variable **y**. The third type is a *free variable*, which is a variable that is neither a formal parameter nor a local variable. An example of a free variable would be **w** in the following function:

```
─────────────────────────────── R ───────────────────────────────
f2 = function(x) {
  return(w*x)
}
```

If you refer to a free variable inside of a function, R looks for its definition in the environment in which the function was defined. Consider the following example, which also demonstrates that you can nest one function inside another:

```
─────────────────────────────── R ───────────────────────────────
f3 = function(x) {
  y = 2
  f4 = function(z) {
    return(y*z)
  }
  return(x + f4(x))
}
```

Inside the function **f4**, there is no local variable **y**. When **y** is referred to, **f4** looks in its parent environment, namely the function **f3**, and uses the value of **y** it finds there. Typically, if you are not writing nested functions, this behavior means that if R cannot find a variable inside of your function, it will use a "global" variable defined at the top-level environment. This behavior is quite different from MATLAB's.

If you assign a value to a free variable, it becomes a local variable [28], rather than affecting a variable in the parent environment. Consider this function:

```
─────────────────────────────── f5.R ───────────────────────────────
1  f5 = function(x) {
2    y = 2
3    f6 = function(z) {
4      tmp = y*z
5      y = 20
6      return(y*tmp)
7    }
8    tmp2 = f6(x)
```

```
9     print(y)
10    return(x + tmp2)
11 }
```

The function **f6** first uses the free variable **y**, obtaining its value of 2 from the parent environment, namely the function **f5**. It then assigns a value to **y**, which creates a local variable within **f6**; the value of **y** within the function **f5** is unchanged, as can be seen by calling `z=f5(3)`, where the printed value of **y** is still 2.

If you wish to assign values to variables in the parent environment of a function, use the `<<-` operator. For example, line 5 in the file **f5.R** above could be changed to y `<<-` 20. This searches up the hierarchy of parent environments until it finds a variable with the specified name, and assigns the given value to it. If it does not find a variable that matches, then a variable by that name is created in the top-level/global environment. If line 5 in **f5.R** is changed as specified above, calling `f5(3)` shows that the value of **y** in the function **f5** does get changed to 20.

If your function has a formal parameter or local variable named **y**, but you still wish to access a top-level/global variable **y**, there is still a way to do so. You can use the expression `get('y', envir=globalenv())`.

8.1.4 Static variables

If you have programmed in a language like C, you may have used something called *static variables*. Normally, if you declare a local variable inside of a function, that variable gets "created" each time the function is called. When the function exits, the variable gets "cleared," i.e., its value is lost. However, you can also create static variables, whose values persist between consecutive calls to the function. Static variables are useful for example if the function needs to create some large data structure, such as a large matrix, which may be time-consuming to initialize. For efficiency, you could have the function set up the data structure the first time it is called, and then re-use it on subsequent calls. Or you may wish the function to open a file the first time it is called, and then send some output to that file each subsequent time it is called. Another common use is to have a counter inside a function which keeps track of how many times the function is called.

R does not provide a way to declare that you wish to make a variable static. Instead, what you can do is create a new local *environment*, essentially a namespace where a collection of variables can reside. This can be done using **local()**. This new local environment will contain a variable (or several variables, if you want more than one static variable), along with your function. The function can then refer to that variable which was created within the local environment. Note that when assigning to the variable, you need to use the `<<-` operator, to assign to the variable in the local environment (but outside of the function) rather than simply assigning to a regular local variable inside of the function. The file below implements this.

———————————— statictest.R ————————————

```
statictest = local({
  y = 0
  myfunc = function(x) {
    y <<- y + 1
    if (y == 1) {
      pluralStr = ''
    } else {
      pluralStr = 's'
    }
```

```
      cat(sprintf('Function called %d time%s', y, pluralStr))
      return(x*x)
   }
})
```

The code below demonstrates how to use it. Note that to effectively clear out the static variable, you can re-**source** the code creating a local environment and the function. The differing outputs of the two **environment** commands below show that rather than clearing the static variable, what is really happening is that a new environment is being created, with a new variable inside it.

———————————————————————— R ————————————————————————

```
> source('statictest.R')
> environment(statictest)
<environment: 0x100d7be38>
> statictest(11)
[1] "Function called 1 time"
[1] 121
> statictest(11)
[1] "Function called 2 times"
[1] 121
> statictest(11)
[1] "Function called 3 times"
[1] 121
> source('statictest.R')
> environment(statictest)
<environment: 0x1219036b0>
> statictest(11)
[1] "Function called 1 time"
[1] 121
> statictest(11)
[1] "Function called 2 times"
[1] 121
```

8.1.5   Variable arguments

Sometimes, it is desirable to write a function which can receive an arbitrary set of extra parameters which are then passed along to some other function. A common example of this is when you write a function which creates, reads, or modifies some data and then plots it. You may wish to pass in some parameters to your function which specify the plotting style. An example is below, which creates **n** random (x, y) points and plots them.

———————————————————————— plotrand.R ————————————————————————

```
plotrand = function(n,...) {
   x = runif(n);   y = runif(n);
   plot(x, y, ...)
}
```

If any extra arguments (beyond the number of points, **n**) are passed to **plotrand**, it simply passes them along to **plot**. So you can call `plotrand(10)` to plot 10 random points with the default plotting style (black circles, with no lines), or use something like the following to specify various attributes of the plot (thick blue lines and large gray filled markers with blue boundaries): `plotrand(10, type='o', col='blue', cex=2, lwd=3, pch=21, bg='gray')`. The expression `...` is a list (in the formal sense, as in Section 5.1) of all of the extra arguments. If for some reason you wanted, say, a list containing only the second and fifth extra parameters, you could use an expression like `list(...)[c(2,5)]`.

8.2 MATLAB

8.2.1 Inline and anonymous functions

MATLAB has two mechanism for creating a function definition which gets stored within a variable, somewhat similar to R. These are called *inline* functions and *anonymous functions*. Do not confuse MATLAB's inline functions with inline functions in a language like C, which are supposed to run more quickly than standard functions. In fact, inline functions in MATLAB run very slowly compared to standard functions defined in files; they should be avoided if you will be calling them very often or performance is a concern. Both methods are described below since you may encounter them, but anonymous functions are now generally the preferred way to handle this type of short function.

Inline functions are constructed from a single MATLAB expression. For example, if you want to define a function which computes the logistic growth rate $f(r, K, x) = rx(1 - x/K)$, you can use the following command:

```
 ———————————— MATLAB ————————————
logistic = inline('r*x*(1-x/K)')
```

Note that the expression is passed to **inline** as a string. MATLAB will determine that the function takes arguments **K**, **r**, and **x**. It decides what order the parameters should go in; if you are not happy with its choice, you can specify the order yourself:

```
 ———————————— MATLAB ————————————
logistic = inline('r*x*(1-x/K)', 'K', 'r', 'x')
```

You can also use `argnames(logistic)` to see the names of the parameters to the function. You can then call your function via a command like `logistic(10,2,4.8)`.

A similar mechanism exists for writing short, single-expression functions in MATLAB, called *anonymous* functions. These run much more quickly than inline functions. To write the logistic function as an anonymous function, you can enter

```
 ———————————— MATLAB ————————————
logistic = @(K,r,x) r*x*(1-x/K)
```

You list the parameters to the function inside the parentheses after the **@**, and then follow that with the expression defining the function. You call this version of **logistic** the same way you called the inline version, e.g., `logistic(10,2,4.8)`.

8.2.2 Writing functions

The standard mechanism for writing a function in MATLAB is to put its code in a file whose name matches the function name (with a ".m" suffix). For example, to write a MATLAB Fibonacci function **fib** analagous to the R version from Section 8.1.1, we would put the following into the file **fib.m**:

```
                                    fib.m
function retval = fib(steps,first,second)
% fib    Compute Fibonacci numbers
%
% FIB(STEPS,FIRST,SECOND)
%   STEPS: how many iterations to do -- if steps=k then compute the
%          (k+1)'st Fibonacci number
%   FIRST,SECOND: the first two numbers in the sequence
%                 (default: 0 and 1, respectively)

% set up default parameters if necessary
if (nargin < 2)
  first = 0;
end
if (nargin < 3)
  second = 1;
end
for i = 1:steps
  nextval = first+second;  % compute next value
  first = second;  % update records
  second = nextval;
end
retval = second;
```

Some things to note about functions in MATLAB:

1. The word **retval** on the first line is not a reserved word; it is simply a variable name I chose. To return a value from the function, store that value in the variable you have specified after the keyword **function** on the first line of your file. (I use **retval**, for "RETurn VALue," merely as a name that is easy for me to remember, and one that I am unlikely to use for anything else.)

2. As mentioned in the previous item, a **return** statement is not used to specify what value the function will return; assigning a value to the special variable you designate in the function definition is used for that. But MATLAB does have a **return** statement, which takes no arguments. It causes execution of your function to finish, and return to the caller.

3. Commands without a semicolon inside a function will display their output, even if you use a semicolon when calling the function. That is, if line 20 of **fib.m** had no semicolon, you would still see its output even if you entered the command **f = fib(20);** with a semicolon. If you are writing functions that someone else may use, it is good etiquette to be sure all lines inside your function have semicolons so that it will run silently.

4. The name of the function on the first line (**fib** in our example here) should match the filename (**fib.m**), although MATLAB currently does not strictly enforce this.

5. Formal parameters are given in parentheses after the function name. There is no way to specify default values for parameters on this line, although there are mechanisms for detecting when parameters are missing (see next item).

6. The special variable **nargin** indicates how many formal parameters were passed in when the function was called. The **fib** function checks this value, and sets

default values for the second and third parameters (**first** and **second**) if they were not specified.

7. To write a function which does not return any values, do not list any variables on the first line. The first line of a file defining such a function would look like **function foo(x,y)**.

8. A function can return multiple values to the caller. To do so, you can simply list the multiple return values on the first line, like this:

```
───────────────────── f1.m ─────────────────────
function [sqr,cube] = f1(x)
sqr = x^2;
cube = x^3;
```

Then just assign values to the return values somewhere within the body of the function. The different return values do not need to have the same data type; for example, you can return a matrix and a string. You can check the special variable **nargout** to see how many return values the caller requested. For example, to return the square and cube of the formal parameter **x** in a single vector if only one or fewer return values were requested by the caller, or in two separate return values, the following function can be used:

```
───────────────────── f2.m ─────────────────────
function [sqr,cube] = f2(x)
sqr = x^2;
cube = x^3;
if (nargout < 2)
   sqr = [sqr cube];
end
```

9. If you put comments immediately below the **function** line, they will be displayed by MATLAB's **help** system. For the **fib** function above, if you type **help fib** (and your search path contains the **fib.m** file), you will see the block of comments at the top of the file displayed. If you have additional comments at the top of your file that you do not want displayed by the help system, put a blank line above the comments you do not want displayed (as with the comment about setting up default parameters in **fib.m**). Also note that the **lookfor** help system will display only the first line of comments after the **function** line if your function (or its first comment line) matches a search.

10. You can put multiple functions within a single file. Subsequent functions within the file (which are entered after the body of the first function) are called *subfunctions*. Subfunctions are only visible within that file (or more precisely, to functions defined within that file); they are not callable from the MATLAB command prompt or by other functions defined in other files. See Section 8.2.3 for an example. You can also define additional functions within other functions (these are called *nested* functions). If you do so, all functions within the file must be terminated with an **end** statement.

8.2.3 Calling functions

Function parameters in MATLAB are specified purely by order. You cannot write something like **fib(first=3)** to specify a parameter by name as is common in R. You also cannot write **f1(y=7)** to assign a value to the variable **y** and then pass that value 7 as a parameter to the function **f1**. Note, however, that as a way around this limitation, some of MATLAB's functions are written to take their parameters in an arbitrary order by using strings to specify which parameter is which. For example, the commands plot(x, y, 'MarkerSize', 10, 'LineWidth', 2) and plot(x, y, 'LineWidth', 2, 'MarkerSize', 10) will behave identically.

Because the function **fib** was written to handle missing values for the second and third parameters,[3] you can call it in a few different ways: **fib(10)** to compute the eleventh Fibonacci number $F_{11} = 89$, or **fib(10,2)** or **fib(10,2,1)** to compute the eleventh Lucas number $L_{11} = 199$.

When a function has multiple return values, you can store the values by calling the function with multiple variables within square brackets. For example, for our function **f1** which computes the square and cube of its argument, you can call **[a,b]=f1(7)** to store the value $7^2 = 49$ in **a** and $7^3 = 343$ in **b**. If you call **c=f1(7)**, only the value 49 will be stored in **c**; although the function itself assigns two return values, the second one is lost if it is not stored. Similarly, calling just **f1(7)** will only display the single value 49.

For the function **f2** which modifies its behavior according to the number of outputs, if zero or one outputs are specified, a vector is returned; if two outputs are specified, they are assigned the two scalar return values. This can be seen in the following sequence of commands and outputs:

```
──────────────────────── MATLAB ────────────────────────
>> f2(7)
ans =
    49    343
>> a=f2(7)
a =
    49    343
>> [a,b]=f2(7)
a =
    49
b =
   343
```

The file below contains a function as well as two subfunctions, to simulate a stochastic birth-death process with constant per capita rates.

```
──────────────────────── birthdeath.m ────────────────────────
function [eventTimes,popSizes] = birthdeath(phi,mu,numEvents,n0)
% birthdeath    Stochastic birth-death process with constant
% per capita rates
eventTimes = zeros(1,numEvents+1);
popSizes = zeros(1,numEvents+1);
eventTimes(1) = 0;  popSizes(1) = n0;
for i=1:numEvents
  if (popSizes(i) == 0)
    eventTimes((i+1):(numEvents+1)) = eventTimes(i);
    popSizes((i+1):(numEvents+1)) = 0;
    break;
  end
  eventTimes(i+1) = eventTimes(i)+interEventTime(phi,mu,popSizes(i));
```

---

[3] You could write a function so that missing parameters are not necessarily assumed to be the last ones. But because of the fact that MATLAB function parameters are specified by position, assuming that missing parameters are the last ones is a good convention to follow.

```
    if (rand < probBirth(phi,mu))
      popSizes(i+1) = popSizes(i) + 1;
    else
      popSizes(i+1) = popSizes(i) - 1;
    end
  end

  % time between consecutive events
  function retval = interEventTime(phi,mu,N)
  totalRate = (phi+mu)*N;
  retval = -log(rand)/totalRate;

  % probability that the next event is a birth
  function retval = probBirth(phi,mu)
  retval = phi/(phi+mu);
```

You could call this function and plot the results by doing, for example, `[et,n]=birthdeath(2, 1, 100, 2); plot(et, n, 'o-')`.

One nice feature of MATLAB is that if you are calling a function without providing any parameters, you can omit the parentheses. That is, you can simply enter **date** to call the **date** function, rather than `date()` as you do in many languages, including R (though you can include the empty parentheses in MATLAB too, if you like). Another convenience built into MATLAB is that you can use a "command-style" format when calling a function which takes only string arguments. For example, if the function **foo** takes two string arguments, then `foo('hi', 'there')` is equivalent to `foo hi there`.

One exception to the above is with anonymous functions that take no arguments. Because a variable referring to an anonymous function is technically a function handle in MATLAB (see Section 8.2.7), if **f** is an anonymous function, typing **f** will simply show you the definition of **f**, rather than calling it. In that case, to call the function you must include empty parentheses, i.e., write `f()`.

8.2.4   Environments and variable scope

In MATLAB, as in R, variables defined within a function are only visible within that function, and they vanish when the function exits (unless they are declared **persistent**). For example, if you define the function

```
────────── f3.m ──────────
function retval = f3(x)
y = 12;
retval = y*x;
```

then the variable **y** has not been set at the top-level or global scope, i.e., if you enter `f1(8)`, then subsequently entering **y** will not show the value 12. When the function **f1** exits, the variable **y** created inside of it disappears. If you already had a variable **y** defined before calling the function, its contents will not be modified. This behavior is common to both R and MATLAB (and many other programming languages as well).

Unlike R, if you try within a function to access a variable which does not exist locally, MATLAB will not examine the parent's workspace to look for the variable. There is a way to share variables between functions, or between a function and the top-level/global/base workspace. Variables can be declared **global**, with a command like `global someVar anotherVar`. If you want two functions to share a variable, they must both declare them global. You can also issue the **global** command at the MATLAB prompt (with accompanying variable names) to share a variable between a function and the base workspace, which is where variables created at the command prompt exist.

A function can also assign a value to a variable in either its parent's (the caller's) or the base (top-level) workspace, via the **assignin** function. You can do, for example, `assignin('base', 'y', 12)` to assign a value to **y** which will be visible at the command prompt, or `assignin('caller', 'y', 12)` to assign a value of **y** in the workspace the current function was called from, i.e., the parent of the function.

If your function has a formal parameter or local variable named **y**, but you still wish to access a top-level/global variable **y**, there is still a way to do so. You can use the expression `evalin('base', 'y')`.

8.2.5   Static variables

A static variable is called a *persistent* variable in MATLAB. You can declare a variable **y** to be persistent via the command **persistent y**, but note that this can only be done within a function defined within a file. A persistent variable is essentially a global variable which is only visible within the function where it is declared. When a variable is first declared as persistent, it is initialized to be an empty (0×0) matrix. You can test for that condition to determine whether the variable needs to be initialized. Consider the following file:

```
——————————————————— statictest.m ———————————————————
function retval = statictest(x)
persistent y
if isempty(y)
  y = 1;  % first time initialization
else
  y = y + 1;  % do this on subsequent calls
end
if (y == 1)
  pluralStr = '';
else
  pluralStr = 's';
end
disp(sprintf('Function called %d time%s', y, pluralStr));
retval = x*x;  % return x^2
```

This behaves as follows:

```
——————————————————— MATLAB ———————————————————
>> statictest(11)
Function called 1 time
ans =
   121
>> statictest(11)
Function called 2 times
ans =
   121
>> statictest(11)
Function called 3 times
ans =
   121
>> clear statictest
```

```
>> statictest(11)
Function called 1 time
ans =
    121
>> statictest(11)
Function called 2 times
ans =
    121
```

The above code shows that you can **clear** the function to remove its static variables from memory, so that they will be reinitialized the next time the function is called.

8.2.6 Variable arguments

Sometimes, it is desirable to write a function which can receive an arbitrary set of extra parameters which are then passed along to some other function. A common example of this is when you write a function which creates, reads, or modifies some data and then plots it. You may wish to pass in some parameters to your function which specify the plotting style. An example is below, which creates **n** random (x, y) points and plots them.

—————————— plotrand.m ——————————
```
function plotrand(n,varargin)
x = rand(n,1);   y = rand(n,1);
plot(x, y, varargin{:})
```

If any extra arguments (beyond the number of points, **n**) are passed to **plotrand**, it simply passes them along to **plot**. So you can call `plotrand(10)` to plot 10 random points with the default plotting style (blue lines, with no point markers), or use something like the following to specify various attributes of the plot (thicker lines and large gray filled markers with blue boundaries): `plotrand(10, 'o-',' LineWidth', 2, 'MarkerSize', 10, 'MarkerFaceColor', [.5 .5 .5])`. The expression **varargin{:}** gets replaced by a comma-separated list of all of the extra arguments. If for some reason you wanted, say, only the second and fifth extra parameters, you could use an expression like **varargin{[2 5]}**.

There is a similar mechanism **varargout** if your function may return an arbitrarily large number of return values (unlike the function **f2** earlier, which returns at most two values). For example, consider a function **moments** which takes a vector **v** of length n as a parameter, and computes $\mu_{(k)} := (1/n) \sum_{i=1}^{n} (v_i)^k$ for $k = 1, 2, 3, \ldots$. The following function will do this:

—————————— moments.m ——————————
```
function varargout=moments(v)
maxk = max(nargout,1);
for k=1:maxk
  varargout(k) = {mean(v.^k)};
end
```

If you then create a vector **v** and do either `moments(v)` or `m = moments(v)`, you only receive the first moment $\mu_{(1)}$ as a return value. But if you call `[a,b,c] = moments(v)`, you receive the first three moments $\mu_{(1)}, \mu_{(2)}$, and $\mu_{(3)}$, which get stored in **a**, **b**, and **c**, respectively. (If you wish to write a function like this in R, the simplest thing to do is simply have the function take an additional argument which indicates how many moments should be computed, and have the function return those moments in a vector or list.)

8.2.7   Function handles

Sometimes it is necessary to pass a function to another function; for example, to minimize a function of one variable, you can use the function **fminbnd**. If you have defined a function **myfunc(x)** that you want to minimize, you need tell **fminbnd** that that is the function you want to minimize. To do so, you use what is called a *function handle*. This is essentially a way of storing a reference to a function within a MATLAB variable. For example, you can do `fh = @sin` to build a handle to the **sin** function. You can then call it like a normal function, e.g., using the command `fh(2)` to compute sin 2. You can also pass it as an argument to other functions, e.g., `fminbnd(fh, 4, 6)`, or more directly, `fminbnd(@sin, 4, 6)`.

When you build an anonymous function as in Section 8.2.1, the variable you create is in fact a function handle. So if you wish to pass it to another function, you do not need to prefix it with an "@."

Also, you can manipulate function handles in ways similar to other variables, except you cannot combine them within a matrix. That is, you cannot do `v = [ @sin @cos]`. There would be confusion between using parentheses to access elements of such a matrix, versus using parentheses to delimit the parameters to the function. You can, however, build a cell array containing function handles:

```
──────────────── MATLAB ────────────────
>> c = {@sin @cos @(x) x^3};
>> c{1}(2)    % call sin(2)
ans =
    0.9093
>> c{3}(8)   % call the anonymous function w/ parameter 8
ans =
   512
```

8.3   Summary of main differences

User-defined functions have some key differences between R and MATLAB, in particular, the way functions are accessed from files, the way parameters are passed to functions, the way multiple values are returned from a function, and how functions search for nonlocal variables. In MATLAB, a function is most commonly written in a file, say **foo.m**. When you type `foo(8)` at the MATLAB command prompt, it goes looking in a specified set of folders (the search path; see Section 7.2) to find the file **foo.m** containing the definition of the function. If you update the file containing the function's code, then the next time you call the function, MATLAB will automatically read and use the updated version from the file. In R, on the other hand, storing a function's code in a file is a convenience. It is up to you to tell R to read the contents of that file to read in the definition of the function. If you update the file, you need to explicitly tell R to read its contents again.

In R, functions can define default values for their parameters. When you call a function, you can specify values for particular parameters in any order, without providing any information about the other parameters. In MATLAB, it is possible to set up default values for function parameters, but with less flexibility. Parameters must be passed to functions in a fixed order. If you want to use default values for the first two parameters but specify a value for the third, you still need to pass in values for those first two parameters. (But note that some functions are written so that if you pass in, say, an empty matrix as a parameter, a default value will be used instead, while other functions use parameter strings to allow you to pass in parameters in an arbitrary order.)

Multiple values can be returned from a MATLAB function and stored in separate variables by the caller. In R, the closest equivalent is for the function to return a list containing the multiple values. And finally, if an R function tries to access a variable that was not defined locally within the function, it will search up the chain of parents (callers), using the value of a variable by that name in one of the ancestors (commonly, the value of a top-level R variable). MATLAB does not automatically search for variables in this way, but you can explicitly request that your function do so.

9

Probability and Random Numbers

Facilities are provided in both packages to work with random numbers following various probability distributions, and to calculate probability densities and cumulative probabilities. In MATLAB, some of the more advanced or powerful functions are part of the Statistics Toolbox, although the core MATLAB does provide the essentials.

Both platforms currently use the Mersenne Twister [20] as the default random number generator, but include other generators as well.

9.1 Basic random values, permutations, and samples

1. Generate a continuous uniform random value between 0 and 1.

R	MATLAB
`runif(1)`	`rand`

Neither function above will generate either of the extreme values 0 or 1, i.e., they both generate values from the open interval (0,1).

2. Generate a vector of n uniform random values between 0 and 1.

R	MATLAB
`runif(n)`	`rand(n,1)` for a column vector, or `rand(1,n)` for a row vector

3. Generate an $m \times n$ matrix of uniform random values between 0 and 1.

R	MATLAB
`matrix(runif(m*n), nrow=m, ncol=n)` or simply `matrix(runif(m*n),m)`	`rand(m,n)`

4. Generate an $m \times n$ matrix of continuous uniform random values between a and b.

R	MATLAB
`matrix(runif(m*n,a,b),m)`	If you have the Statistics Toolbox: `unifrnd(a,b,m,n)` Otherwise: `a+rand(m,n)*(b-a)`

5. Generate a random integer between 1 and k from the discrete uniform distribution.

R	MATLAB
`sample(k,1)`	`randi(k)`

6. Generate an $m \times n$ matrix of discrete uniform values between 1 and k.

R	MATLAB
`matrix(sample(k, m*n, replace=TRUE), m)`	`randi(k, m, n)`

7. Generate an $m \times n$ matrix where each value is 1 with probability p, and otherwise 0.

`matrix(sample(c(0,1), m*n,` `replace=TRUE, prob=c(1-p,p)), m)` Or: `matrix(runif(m*n)<p, m)*1`	`(rand(m,n)<p)*1`

The "***1**" above converts logical values back into numeric values.

8. Generate an $m \times n$ matrix where each value is **a** with probability p, and otherwise **b**.

`matrix(sample(c(b,a), m*n,` `replace=TRUE, prob=c(1-p,p)), m)` Or: `b+(a-b)matrix(runif(m*n)<p, m)`	`b+(a-b)*(rand(m,n)<p)`

9. Generate a random integer between **a** and **b** inclusive from the discrete uniform distribution.

`a+floor((b-a+1)*runif())` or `sample(a:b, 1)`	If you have the Statistics Toolbox: `unidrnd(b-a+1)+a-1` Otherwise: `a+floor((b-a+1)*rand)` or `a-1+randi(b-a+1)`

10. With probability p, perform a given set of commands.

`if (runif(1) < p) {` `   ...some commands...` `}` Or: `if (sample(c(TRUE,FALSE),1,` `prob=c(p,1-p))) {` `   ...some commands...` `}`	`if (rand < p)` `   ...some commands...` `end`

11. Generate a random permutation of the integers from 1 to **n**.

`sample(n)`	`randperm(n)`

12. Sample **k** values between 1 and **n** without replacement.

`sample(n, k)`	If you have the Statistics Toolbox: `randsample(n, k)` Otherwise: `ri=randperm(n); ri=ri(1:k)`

13. Choose **k** values from the vector **v** (with replacement), storing the results in **w**.

`w=sample(v, k, replace=TRUE)`	If you have the Statistics Toolbox: `w=randsample(v,k,true)` Otherwise: `w=v(floor(length(v)*rand(k,1))+1)`

Both the **sample** and **randsample** commands above will have trouble if the vector **v** may possibly contain only a single value. If **v** contains only the single value **n**, then both commands will sample from the integers between 1 and **n**, rather than just choosing the actual value **n** itself. To get around this problem, you can instead first generate **k** integers between 1 and **length(v)**, and then use those as indices into **v**, as follows:

`inds = sample(length(v), k)` `w = v[inds]`	`inds = randsample(length(v),k,true)` `w = v(inds)`

14. Choose **k** values from the vector **v** (without replacement), storing the results in **w**.

w=sample(v, k)	If you have the Statistics Toolbox: w=randsample(v,k)　　Otherwise: inds=randperm(length(v)); w=v(inds(1:k))

15. Generate a random integer between 1 and **n**, with corresponding probabilities in vector **pv**.

sample(n,1,prob=pv)	If you have the Statistics Toolbox: randsample(n,1,true,pv)　　Otherwise: sum(rand > cumsum(pv)/sum(pv))+1

(MATLAB): If you know the entries of **pv** sum to 1, you can omit the /**sum(pv)** in the second form above.

9.2　Random number seed

When performing stochastic simulations, or any calculations involving random numbers, it is sometimes essential to be able to reproduce particular results. This can be done by using a given seed value to set the random number generator back to a known state, so that the sequence of random numbers subsequently produced will be reproducible by using the same seed again. Although the internal state of the Mersenne Twister is fairly complex (a vector containing more than 600 integers), you can use a single integer as a seed in both platforms. To set the seed to the value 12, for example, in R use set.seed(12), and in MATLAB use rng(12)[1].

By default, R seeds the random number generator based on the current time and process ID the first time you generate a random value. MATLAB begins with the same seed each time; you can reset the random number seed back to this initial state via the command rng('default'). You can use the current time to seed the random number generator in MATLAB via the command rng('shuffle').

You may also wish to make your results repeatable without using a specified fixed random number seed. The simplest way to do this is to first generate a random integer, and use that as a random number seed before proceeding. For example, to use (and save for later re-use) a random seed between 1 and 10^9, you could do the following:

R	MATLAB
saved.seed = sample(1e9,1) set.seed(saved.seed)	saved.seed = randi(1e9); rng(saved.seed)

---

[1]In older versions of MATLAB, you would instead use rand('state',12). Using this command in a newer version of MATLAB activates a legacy random number generator; use rand('default') to switch back to the standard one.

9.3   Random variates from probability distributions

Here, methods are introduced for generating random values from specified distributions, such as normal, Poisson, exponential, and so on. One of the main differences between the functions for generating random values is that R's functions use the first parameter to indicate how many random values you want (in a vector), followed by the parameters for the distribution; MATLAB's functions take the parameters for the distribution first, followed by parameters indicating how many rows and columns in the matrix of random values (i.e., how many random values you want).

In R, the various functions to generate random values all begin with the letter "r." Each command below generates **w** independent, identically distributed values from the given probability distribution. If you want an $r \times c$ matrix of such values, use a command such as `matrix(rbinom(r*c,n,p),r)`. Use the command `?Distributions` to see a list of the standard probability distributions available.

In MATLAB, the various functions to generate random values all end with the letters "rnd." Each command below generates an $r \times c$ matrix of independent, identically distributed values from the given probability distribution. An alternative approach is the **random** function, which takes the name of a distribution, followed by parameters, and optionally a set of sizes. For example, to generate a 4×8 matrix of random values from the normal distribution with mean 15 and standard deviation 16, you can use `random('norm', 15, 16, 4, 8)`. Note that these functions are part of MATLAB's Statistics Toolbox, although alternatives are available in some cases for those without that additional Toolbox.

16. Generate values from the binomial distribution with parameters **n** and **p**.

R	MATLAB
`rbinom(w,n,p)`	`binornd(n,p,r,c)`

(MATLAB): If you do not have the Statistics Toolbox, you can instead use the command `reshape(sum(rand(n,r*c) < p), r, c)`. This sums over Bernoulli trials to generate binomial values, and then reshapes the results into an $r \times c$ matrix.

17. Generate values from the Poisson distribution with mean λ.

R	MATLAB
`rpois(w,lambda)`	`poissrnd(lambda,r,c)`

18. Generate values from the exponential distribution with mean μ.

R	MATLAB
`rexp(w,1/mu)`	`exprnd(mu,r,c)`

Note that R's **rexp** function takes the reciprocal of the mean as its parameter to characterize the probability distribution, while MATLAB's **exprnd** takes the mean itself.

(MATLAB): If you do not have the Statistics Toolbox, you can instead use the command `-mu*log(rand(r,c))`.

19. Generate values from the discrete uniform distribution on integers $1, \ldots, k$.

R	MATLAB
`sample(k,w,replace=TRUE)`	`unidrnd(k,r,c)`

(MATLAB): Without the Statistics Toolbox, use: `randi(k,r,c)`.

20. Generate values from the continuous uniform distribution on the interval (a, b).

R	MATLAB
`runif(w,a,b)`	`unifrnd(a,b,r,c)`

(MATLAB): Without the Statistics Toolbox: `a+(b-a)*rand(r,c)`.

21. Generate values from the normal distribution with mean μ and standard deviation σ.

rnorm(w,mu,sigma)	normrnd(mu,sigma,r,c)

(MATLAB): Without the Statistics Toolbox: `mu + sigma*randn(r,c)`.

22. Generate values from the chi-squared (χ^2) distribution with **df** degrees of freedom.

rchisq(w,df)	chi2rnd(df,r,c)

(MATLAB): Without the Statistics Toolbox: `sum(randn(r,c,df).^2,3)`. This repeatedly sums the squares of sets of **df** values from the standard normal distribution.

23. Generate values from the noncentral chi-squared distribution with **df** degrees of freedom and noncentrality parameter **delta**.

rchisq(w,df,delta)	ncx2rnd(df,delta,r,c)

24. Generate values from the F distribution with parameters **d1** and **d2**.

rf(w,d1,d2)	frnd(d1,d2,r,c)

25. Generate values from the noncentral F distribution with parameters **d1** and **d2** and noncentrality parameter **delta**.

rf(w,d1,d2, delta)	ncfrnd(d1,d2,delta,r,c)

26. Generate values from the gamma distribution with shape parameter **k** and scale parameter θ.

rgamma(w,k,theta)	gamrnd(k,theta,r,c)

(R): Alternatively, you can use `rgamma(w, alpha, rate=beta)` to produce values using shape parameter α and rate parameter $\beta = 1/\theta$.

27. Generate values from the geometric distribution with parameter **p**.

rgeom(w,p)	geornd(p,r,c)

28. Generate values from the hypergeometric distribution counting the number of white balls seen when drawing **n** balls from an urn containing a total of **m** balls, **k** of which are white.

rhyper(w,k,m-k,n)	hygernd(m,k,n,r,c)

(R): The three parameters to the **rhyper** function are the number of white balls in the urn, the number of black balls in the urn, and the number of balls drawn from the urn, which is different from MATLAB's **hygernd** function.

29. Generate values from the lognormal distribution with parameters μ and σ being the mean and standard deviation of the logarithm of the values.

rlnorm(w,mu,sigma)	lognrnd(mu,sigma,r,c)

30. Generate values from the negative binomial distribution with parameters **n** and **p**.

rnbinom(w,n,p)	nbinrnd(n,p,r,c)

31. Generate values from the beta distribution with parameters **a** and **b**.

rbeta(w,a,b)	betarnd(a,b,r,c)

32. Generate values from the Student t distribution with **df** degrees of freedom.

rt(w,df)	trnd(df,r,c)

33. Generate values from the noncentral Student t distribution with **df** degrees of freedom and noncentrality parameter **delta**.

rt(w,df,delta)	nctrnd(df,delta,r,c)

34. Generate values from the Weibull distribution with shape parameter **b** and scale parameter **a**.

rweibull(w,b,a)	wblrnd(a,b,r,c)

   R and MATLAB give the two parameters **a** and **b** to these routines in opposite order.

35. Generate random vectors from the multinomial distribution with **n** trials and probability vector **p**.

rmultinom(w,n,p)	mrnd(n,p,r)

   (R): This generates **w** random vectors, and returns those vectors as the *columns* within a matrix.

   (MATLAB): This generates **r** random vectors, and returns those vectors as the *rows* within a matrix.

9.4   PDFs, CDFs, and inverse CDFs

Both platforms also have function to compute PDFs (probability density functions), CDFs (cumulative distribution functions), and inverse CDFs. Again, in MATLAB, these functions are part of the Statistics Toolbox.

R

There is a PDF function corresponding to each of the functions that generates random values. It can be obtained by changing the "r" (for "random") at the beginning of the function name to "d" (for "density"). The first argument to the PDF functions is the value at which you want to evaluate the function; the later arguments are the parameters of the probability distribution. For example, to compute the density at $x = 0.2$ for a normal distribution with mean 1.3 and standard deviation 1.1, use dnorm(0.2, 1.3, 1.1). Note that R does not include functions for the discrete uniform distribution, as the **sample** command can be used to generate values from that distribution (see Item 9 in Section 9.1). However, you can piece together the discrete uniform PDF (or more precisely, probability mass function) yourself. For example, the PDF for the discrete uniform distribution on the values $1 \ldots n$ can be computed via the following function.

```
————————————————————————— R —————————————————————————
# PDF for value x from discrete uniform distribution on values 1...n
dunid = function(x, n) {
  return((((x==round(x)) & (x >= 1) & (x <=n))/n)
}
```

Similarly, there is a CDF function corresponding to each random variate generator function. It can be obtained by changing the initial "r" in the function name to "p" (for "probability"). The first argument to the CDF functions is the value x for which you want to evaluate the cumulative probability $P(X \leq x)$. For example, to compute the cumulative probability $P(X \leq 1.3)$ for a random value X following an exponential distribution with mean 0.7, use `pexp(1.3, 1/0.7)`. A function implementing the CDF for the discrete uniform distribution on values $1 \ldots n$ is below.

—————————————— R ——————————————

```
# CDF for value x from discrete uniform distribution on values 1..n
punid = function(x, n) {
 return((x >= 1)*pmin(n,floor(x))/n)
}
```

Finally, the inverse CDF functions can be obtained by changing the initial letter in the function name to "q" (for "quantile"). The first argument to the quantile functions is the probability p; the function then finds the value x satisfying $P(X \leq x) = p$. For example, `qexp(0.7, 1/3)`. A function implementing an inverse CDF for the discrete uniform distribution on values $1 \ldots n$ which behaves identically with MATLAB's is below.

—————————————— R ——————————————

```
# inverse CDF for value x from discrete uniform distribution on values 1..n
qunid = function(p, n) {
   return(ifelse((p<=0)|(p>=1),NaN,ceiling(p*n)))
}
```

MATLAB

There is a PDF function corresponding to each of the functions that generates random values. It can be obtained by changing the "rnd" (for "random") at the end of the function name to "pdf." The first argument to the PDF functions is the value at which you want to evaluate the function; the later arguments are the parameters of the probability distribution. For example, to compute the density at $x = 0.2$ for a normal distribution with mean 1.3 and standard deviation 1.1, use `normpdf(0.2, 1.3, 1.1)`.

Similarly, there is a CDF function corresponding to each random variate generator function. It can be obtained by changing the suffix "rnd" in the function name to "cdf." The first argument to the CDF functions is the value x for which you want to evaluate the cumulative probability $P(X \leq x)$. For example, to compute the cumulative probability $P(X \leq 1.3)$ for a random value X following an exponential distribution with mean 0.7, use `expcdf(1.3, 0.7)`.

Finally, the inverse CDF functions can be obtained by changing the suffix in the function name to "inv." The first argument to the inverse CDF functions is the probability p; the function then finds the value x satisfying $P(X \leq x) = p$. For example, `expinv(0.7, 3)`.

10

Graphics

Both R and MATLAB include facilities for producing a wide variety of high-quality graphics. Both platforms have fairly straightforward and similar ways of producing simple scatterplots of 2-D data, with a large set of additional routines to modify those plots or produce other types of graphical output.

R has two fundamentally different ways of producing graphics (what are called graphics systems): the traditional graphics system and the **grid** graphics system. The two systems do not interact very well, so it is best to try and work with just one system at a time. The grid graphics system is in some ways more powerful and convenient than the traditional graphics system, depending on what you are doing. Some things, such as figure legends, margins, etc. have better defaults or automatic behavior in grid graphics than in traditional graphics. Personally, when I was getting started in R, I found the traditional graphics system to be more similar to the MATLAB graphics commands I was used to, and so I primarily use traditional graphics in R, with occasional exceptions. Paul Murrell has written an excellent reference [21] covering traditional and grid graphics, which is an essential reading for R users.

10.1 Creating, selecting, and closing figure windows

In both R and MATLAB, you can have multiple figure windows open simultaneously.[1] Only one window is considered "active" at a time, where subsequent plotting commands will display their results.

10.1.1 Creating windows

In both environments, trying to plot something will automatically create a new figure window if none exists. However, it is often nice to create additional ones, so that multiple figures may be displayed simultaneously.

To create a new figure window, in R you can use the command `dev.new()`. There are also device-specific commands: in Windows, you can use `windows()`, in Mac OS-X you can use `quartz()`, and in Linux you can use `x11()`.[2]

In MATLAB, simply enter the command `figure` to create a new figure window.

---

[1]Currently, RStudio is an exception to this. It lets you cycle through different figures within a single subwindow in its interface.

[2]You can also use `x11()` under Mac OS-X if you prefer to have graphics appear in an X11 window. And you can use `options(device='x11')` or `options(device='quartz')` to specify the default device that you want `dev.new()` to create. In RStudio, if you really want multiple figure windows, `dev.new()` will not work, but you can use one of the device-specific commands to create a separate figure window.

10.1.2   Listing and selecting windows

In both environments, figure windows can be identified via integers, referred to as device numbers (R) or figure numbers (MATLAB). Figure number 1 is reserved in R as a "null device," with active figure windows beginning with figure number 2. In MATLAB, figure numbers begin with 1. To get a list of open figure windows, in R, use `dev.list()`. In MATLAB, use `get(0,'children')`.

To select figure number **n** and make it the active one, in R, use `dev.set(n)`. In Mac OS, you can also click on a Quartz figure window, and then select Window → Activate Quartz Device Window from the menu. In MATLAB, use `figure(n)`, which will create the figure if one does not already exist with that number. Alternatively, simply clicking in a MATLAB figure window will make it the active one.

To find out which figure number is the currently active one, in R use `dev.cur()`; in MATLAB use `gcf`. R will also give an indication of which figure window is active; the active one will have the word ACTIVE or an asterisk [*] in the titlebar, depending on the operating system and which type of graphics device the window is using (e.g., Quartz or X11).

10.1.3   Closing windows

To close a figure window, you can use your platform's standard method for closing windows (typically, clicking on some type of "x" or red button in the window's title bar, or using a keyboard shortcut such as Cmd-W in Mac OS). To do it via commands, in R use `dev.off()` to close the current figure window, `dev.off(n)` to close figure number **n**, and `graphics.off()` to close all figures. In MATLAB, the corresponding commands are `close`, `close n`, and `close all`.

10.2   Basic 2-D scatterplots

One of the simplest and perhaps most common type of graphics is a scatterplot of (x, y) data. Say we have vectors **x** and **y** of the same length. Both platforms allow you to simply type `plot(x,y)`, although doing so produces different types of plots (just points in R, and just lines in MATLAB). When producing 2-D scatterplots, the options you typically want to specify are: (1) color; (2) plotting symbols, if any, placed at the data points; and (3) the type of lines, if any, connecting the points.

R

The color used in a plot can be specified via a **col** parameter to the **plot** function. The color itself can be specified in a couple of different ways. First, you can use a named color, e.g., **col='red'**. Enter the command `colors()` to see a list of available named colors (there are 657 on my system). Second, you can use Red-Green-Blue hexadecimal triplets. For example, **col='#FF00FF'** gives red=255, green=0, and blue=255, which corresponds to the color magenta. You can use **col2rgb** to see the RGB triplet for a named color, e.g., `col2rgb('magenta')`. You may also find the **rgb** function useful to construct colors from numerical values, e.g., `rgb(1,0,1)` will build the string **'#FF00FF'**.

As for what is actually plotted, in R, you do not simply specify plotting symbols and line types as you do in MATLAB. Instead, the **plot** command can take a parameter **type** that indicates which overall kind of plot to make. Each type is specified using a single character,

'p'	Points
'l'	Lines (that is a lowercase "L", not the number 1)
'b'	Both points and lines, but there are small gaps in the lines at each data point to make room for a plotting symbol
'c'	Only the lines from type **'b'**, with no symbols at the points
'o'	"Overplotted," both lines and points but with no gaps in the lines around the points
'h'	Vertical lines like a histogram
's'	Steps that move first horizontally then vertically
'S'	Steps that move first vertically then horizontally
'n'	No plotting (can be used to set up axes, plot titles, etc.)

TABLE 10.1
Possible values for the **type** parameter to the R **plot** command.

'a'	The character "a". Other characters in quotes can be similarly used.
0	Open square
1	Open circle
2	Triangle pointing up
3	+ (plus)
4	× (cross)
5	Diamond
6	Triangle pointing down
(numbers 7–25)	Various other shapes
'.'	Rectangle of size 0.01 inch, 1 pixel, or 1 point (1/72 inch) depending on device

TABLE 10.2
Values for the **pch** ("Plotting CHaracter") parameter to the R **plot** command.

as listed in Table 10.1. For example, `plot(x, y, type='o')` plots data using both lines and points.

You can then use the **pch** ("Plotting CHaracter") and **lty** ("Line TYpe") parameters to specify the plotting symbol and line type to use, as listed in Tables 10.2 and 10.3. Note that for the line type, you can specify either an integer from the first column of Table 10.3 (e.g., **lty=2**), or a character string from the second column (e.g., **lty='dashed'**).

For example, `plot(x, y, col='red', type='o', lty=2)` plots using red circles and dashed lines, while `plot(x, y, type='p', pch=0)` plots using black squares with no lines. To plot using a blue dash-dot line with no markers, use `plot(x, y, type='l', lty=4)`. To plot using small circles as the markers with a solid line, use `plot(x, y, type='o', pch=16)`. Various other parameters can be used to give further control over things like line width and plotting character colors and size; see the help for **plot**, **points**, and **lines** for more, as well as Section 10.13.

MATLAB

When making a basic plot in MATLAB, you can provide a single character string specifying the color, plot marker or symbol, and line style. It is best to put them in that order, as explained below. There are 8 colors, 13 plot markers, and 4 line styles you can choose from,

0	blank
1	solid
2	dashed
3	dotted
4	dotdash
5	longdash
6	twodash

TABLE 10.3
Values for the **lty** ("Line TYpe") parameter to the R **plot** command. Integers from the first column can be used (e.g., **lty=2**), or strings from the second column (**lty='dashed'**).

Symbol	Color	Symbol	Marker	Symbol	Line style
b	blue	.	point (.)	–	solid line
g	green	o	circle (o)	:	dotted line
r	red	x	cross (×)	-.	dash-dot line
c	cyan	+	plus sign (+)	--	dashed line
m	magenta	*	asterisk (∗)		
y	yellow	s	square (□)		
k	black	d	diamond (◊)		
w	white	v	triangle (down) (▽)		
		^	triangle (up) (△)		
		<	triangle (left) (◁)		
		>	triangle (right) (▷)		
		p	pentragram star		
		h	hexagram star		

TABLE 10.4
Available colors, plotting symbols/markers, and line styles for basic MATLAB plots.

as shown in Table 10.4. You can optionally choose an entry from each column in the table. If you do not specify a color, blue is used as the default. If you specify a plotting symbol but no line type, then no lines are drawn; similarly if you specify a line style but no plotting symbol. If you do not specify any kind of plotting style, blue lines are used.

For example, `plot(x, y, 'ro--')` plots using red circles and dashed lines, while `plot(x, y, 'ks')` uses black squares with no lines. To plot using a dash-dot line, with the default color of blue and no markers, use `plot(x,y,'-.')`. To plot using small circles as the markers with a solid line, use `plot(x,y,'.-')`. To avoid the obvious confusion between these last two, it is best to always specify the color, marker, and line style in that order. Various other parameters can be passed along to **plot** to control things like line width, and the colors and size of markers; see the end of the **plot** help for more, as well as Section 10.13.

Note that internally, MATLAB keeps "handles" to various things plotted in a figure. You can save the handle when adding something to a figure, and then get and set various properties (via functions with those names). That is, the commands `h = plot(x, y);` `get(h)` will display the list of properties of the plot that you can set. You can then change those properties via commands like `set(h, 'LineWidth', 2)`.

10.3   Adding additional plots to a figure

Sometimes, after plotting some data, you wish to add additional plots within the same figure window. It is fairly straightforward to do this in both platforms, but there are some key differences between the two.

R

The **points** and **lines** functions both behave like **plot**, and take the same parameters as **plot**, but they add the new plots to an existing figure rather than replacing what was there. The only difference between **points** and **lines** is that the former defaults to **type='p'** while the latter defaults to **type='l'**. So points(x2, y2) will add some points to an existing plot, while lines(x2, y2) will add lines. You can in fact use lines(x2, y2, type='p').

One major limitation of adding things to plots in this way is that, unlike MATLAB, R will not automatically adjust the axis ranges for you to ensure that all of the old and new data are visible. Once the axis ranges are set up by the initial plot when using traditional graphics, they are fixed. You should therefore make sure those ranges will make all data that are to be plotted visible when your plot is completed; see Section 10.4 for how to specify axis ranges.

MATLAB

In MATLAB, if you enter the command `hold on`, then all subsequent plotting commands will add their graphics to an existing plot, rather than replacing what was there. Coordinate axes will automatically update their ranges so that all data from the existing plots as well as the new one are visible. You can enter `hold off` to turn this off, so that the next plotting command will overwrite the current figure.

10.4   Axis ranges

You can specify the coordinate ranges for the axes in a figure, if you wish to override the defaults. In R, this is necessary if you plan to add additional plots to a figure and the secondary data lie outside the range of the first data you plot. In both platforms, you may wish to adjust the axis ranges simply to make your plots look better, or to leave room for a figure legend.

In R, you can use the **xlim** and **ylim** parameters to the **plot** function to specify the ranges for the x and y axes, respectively. Each parameter value should be a vector of length 2, with the minimum and maximum values. For example, plot(x, y, xlim=c(0,17), ylim=c(-5,42)). Note that R expands the ranges of x and y values by 4% on each end, and then further adjusts things slightly to make the axes look better. You can eliminate this 4% padding in the x and y directions by including the parameters **xaxs='i'**, **yaxs='i'** in your **plot** command; see the help for **par** for more information.

In MATLAB, you can adjust the axis ranges any time after making a plot, via the **axis** command. You provide a vector of length 4, containing the min and max x values, and min and max y values in that order. For example, axis([0 17 -5 42]). The nice thing about this feature is that you can make repeated adjustments to the axis ranges until you have gotten things looking just the way you like.

10.5   Logarithmic axis scales

In R, the **log** parameter to the **plot** function specifies which axes you want to have logarithmic scales. Specifically, you can use **log='x'**, **log='y'**, and **log='xy'** to use logarithmic scales for the x axis, the y axis, and both axes, respectively.

In MATLAB, three different functions are provided to achieve the above results: **semilogx**, **semilogy**, and **loglog**. These functions behave just like **plot**, but with the different scales for the axes.

10.6   Background grid

Sometimes it is useful to add a grid to the background of your plot, to make it easier to identify the coordinates of displayed points. In R, the command `grid()` will add a grid with gray dashed lines. The parameters **nx** and **ny** let you specify how many lines you want in the x and y directions, while parameters **col**, **lty** and **lwd** let you specify the color, line type, and line width of the lines. These are especially useful, since the default line style is not very visible on some devices.

In MATLAB, you can use `grid on` to turn on a grid, `grid minor` to turn on a finer one, and `grid off` to remove the grid.

10.7   Plotting multiple data sets simultaneously

Section 10.3 describes how to include additional plots within an already-existing figure, but sometimes it is more convenient to plot several things at once. For example, say you have vectors **x1**, **y1**, **x2**, and **y2** containing two sets of (x, y) coordinates, and wish to plot two separate curves displaying the data.

R

R has a couple of mechanisms for doing this. First, **matplot** lets you provide matrices **X** and **Y** whose columns contain x and y coordinates. You can bind the vectors into columns of the needed matrices using **cbind**, e.g., with a command like `matplot(cbind(x1, x2), cbind(y1, y2))`. The disadvantage of this approach is that it requires the different curves to have the same length, i.e., the vectors **x1** and **x2** should be the same length, as should **y1** and **y2**. If the original vectors have different lengths, you can pad the shorter ones with enough **NA** values to match the length of the longest one. For example, if the second data set has three fewer points than the first one, this command can be used: `matplot(cbind(x1, c(x2, rep(NA, 3))), cbind(y1, c(y2, rep(NA, 3))))`. You can provide separate plotting parameters (e.g., plot type, color, plotting character, line type) using vector ver-

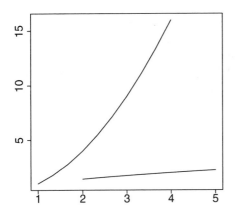 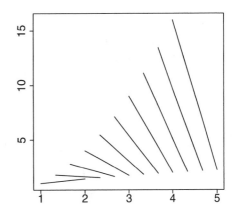

FIGURE 10.1
Left: a plot of matrix data produced by `matplot(X, Y)` (in R) or `plot(X, Y)` (in MAT-LAB). Right: corresponding image of transposed data produced by `matplot(t(X), t(Y))` or `plot(X', Y')`.

sions of the standard **plot** parameters. For example, `matplot(X, Y, type=c('o', 'b')`, `col=c('red', 'blue'), lty=c(2, 4))`. The **type** parameter can be specified more compactly by packing the elements of the vector into a single string, e.g., **type='ob'**. Also, these parameter vectors will be recycled if necessary (see Page 23). There are separate **matpoints** and **matlines** functions which behave like **matplot** but which add their new graphics to an existing plot, just as **points** and **lines** do. Note that transposing the data matrices will produce very different-looking plots. For example, for ten equally spaced x_1 values between 1 and 4 and x_2 values between 2 and 5, with corresponding y values satisfying $y_1 = x_1^2$ and $y_2 = \sqrt{x_2}$, respectively, Figure 10.1 shows the plots that result when using **matplot** with original data and a transposed version of the data. The figures were generated via the following commands:

```
                                  R
x1 = seq(1, 4, len=10);   x2 = seq(2, 5, len=10)
y1 = x1^2;   y2 = sqrt(x2)
X = cbind(x1, x2);   Y = cbind(y1, y2)
matplot(X, Y, type='l', col='black', lty=1, xlab='', ylab='', lwd=2)
dev.new()
matplot(t(X), t(Y), type='l', col='black', lty=1, xlab='', ylab='', lwd=2)
```

The second way to plot multiple data sets simultaneously in R is by combining the separate x vectors into a single one, with **NA** values separating the coordinates corresponding to different data. Each **NA** in the x and y vectors causes R to "pick up the pen" before moving on to the next points. So for example, `plot(c(x1, NA, x2), c(y1, NA, y2), type='o')` will plot two separate curves. The limitation of this approach is that you cannot specify different parameters (plot type, plotting character, line type, color) for the different curves.

MATLAB

MATLAB also has a couple of ways to simultaneously plot multiple data sets. First, MATLAB's **plot** function includes functionality similar to R's **matplot**. The command `plot(X, Y)` plots each column of **Y** against the corresponding column of **X**. You cannot specify plotting parameters separately for each curve; by default, differently colored lines are used. If you instead enter `plot(X, Y, 'o')`, differently colored points are used. Transposing data matrices will produce plots that look quite different. See Figure 10.1 for images of original data and a transposed version of the data, produced by the following commands:

```MATLAB
x1 = linspace(1, 4, 10);  x2 = linspace(2, 5, 10);
y1 = x1.^2;  y2 = sqrt(x2);
X = [x1' x2'];  Y = [y1', y2'];
plot(X, Y, 'k-');  figure;  plot(X', Y', 'k-')
```

MATLAB also allows you to include multiple $(x, y,$ style) triplets in a **plot** command. For example, `plot(x1, y1, 'ro-', x2, y2, 'k*')` will plot the first data set with red circles and solid lines and the second data set with black stars only.

10.8   Axis labels and figure titles

Unlike MATLAB, by default R labels the x and y axes of plots according to the data used to generate them. For example, if you use the command `plot(A[,3], yvec)`, then the x axis will be labeled **A[,3]** and the y axis will be labeled **yvec**. As in MATLAB, there will be no title on the plot by default. You can specify labels for the x and y axes in an R plot, along with a title, by using the **xlab**, **ylab**, or **main** parameters to the **plot** function. For example, `plot(xdat, ydat, xlab='time', ylab='humidity', main='moisture versus time')`. You can also use the **sub** parameter to specify a subtitle at the bottom of the plot. If you wish to instead add the labels after making the plot, you can use the **title** function, which accepts all of the above arguments. However, note that **title** adds the labels you specify, rather than replacing them. If you use `title(xlab='days elapsed')` and there was already a label on the x axis (such as the default one R provided), the new label's text will appear on top of the existing one. This is in contrast with MATLAB, which replaces any existing axis label when you provide a new one. If you wish to add labels later, you may wish to use empty strings or the value **NA** for the labels when first making the plot, using arguments **xlab=NA, ylab=NA** to suppress the default labels.

MATLAB has separate functions **xlabel**, **ylabel**, and **title** for creating axis labels and a figure title. You can call them after using a **plot** command to create the figure, as in the following sequence of commands: `plot(x, y); xlabel('time'); ylabel('humidity'); title('moisture versus time')`. Calling these functions to set an axis label or the title replaces any previous label that may have been present.

10.9   Adding text to figures

Both platforms allow you to add text labels to plots via the command `text(x, y, 'my string')`.

In R, by default, the text is centered at the specified coordinates. You can use the **adj** parameter with a vector containing two elements, each within the range from 0 to 1, to specify adjustments in both the x and y directions. The value 0 specifies the left or bottom; 0.5 specifies the middle, and 1 specifies the right or top. To have the left end of the text at the specified coordinates, but still centered in the vertical direction, use `text(x, y, 'my string', adj=c(0, 0.5)`, and so on. You can add multiple strings simultaneously by passing in vectors for the coordinates and strings, e.g., `text(c(4,8), c(15,16), c('label1', 'label2'))`.

In MATLAB, by default, the left end of the text is at the specified coordinates, and the text is vertically centered. You can use a command like `text(x, y, 'my string', 'HorizontalAlignment', 'center', 'VerticalAlignment', 'baseline')` to horizontally center the text at the given coordinates (use **'right'** to position the right end of the text at the specified coordinates) and raise the text so its baseline is at the specified y coordinate. You can add multiple strings simultaneously by passing in vectors for **x** and **y**, and an array of strings which has the same number of rows as the lengths of the coordinate vectors. However, note that all rows of the array must have the same length, so you will need to pad shorter strings with spaces. For example, you could use a command like `text([4 8], [15 16], 'short   ', 'long label')`. If you are centering the text, the added spaces will mess up the positioning; in that case, it is better to add the labels one at a time. Use the command `h = text(x, y, 'my string')` to save a handle to some text, followed by `set(h)` to see a list of object properties you can set with text strings.

10.10   Greek letters and mathematical symbols

Both platforms allow you to include mathematical expressions, including Greek letters, in text annotations such as axes labels and figure titles. Let us say we wish to plot some (x, y) data with a title of "$y = \sqrt[3]{x^7}$," a y-axis label of "$\lambda = \frac{\phi}{3.5}$ (but where the value 3.5 is obtained from the variable **w**, rather than hardcoded in the code), and with an added text label of "$E[\xi]$" at position $(0.5, 0.6)$.

R

See the help for **plotmath** for information about the variety of mathematical symbols available. To build the plot with the desired main title and y-axis label, the following commands can be used:

```
                                    R
plot(x, y, main=expression(y == sqrt(x^7,3)),
    ylab=bquote(lambda == over(phi,.(w))))
```

The main title is not too bad: we use **expression** to build a mathematical expression; a double-equals == is displayed as a single equals sign in the output, **sqrt(a,b)** is used to produce $\sqrt[b]{a}$, and **a^b** is used to produce a^b. The y-axis label is more complicated, because we wish to substitute in the value from the variable **w**. If we simply wrote **ylab = expression(lambda == over(phi, w))**, we would get $\lambda = \frac{\phi}{w}$, i.e., we would see **w** rather than the value contained in **w**. The function **bquote** works much like **expression** here, except things surrounded by **.( )** are evaluated in the parent frame by default, which substitutes in the value for **w**.

To add the text label, if we just write **text(0.5, 0.6, expression(E[xi]))**, we will instead get E_ξ, because square brackets are used to surround subscripts. The way around this is to use **paste** to combine pieces of the expression, to avoid the square brackets triggering a subscript: text(0.6, 0.4, expression(paste('E[', xi, ']'))). Obviously, these things can get quite complicated fairly quickly, and it is better to tweak commands like this in a script rather than trying to get them right by typing them directly in the R console.

MATLAB

If you know LaTeX, the great news is that MATLAB allows you to include some TeX/LaTeX in your labels. (If you do not, there is an abundance of introductions/tutorials and references about basic LaTeX mathematical expressions available on-line.) Depending on what you are doing, you need to surround your expression with dollar signs to ensure LaTeX is in math mode, and you may also need to explicitly request MATLAB to use LaTeX to interpret your labels (rather than TeX). We can use the following commands:

```
————————————————— MATLAB —————————————————
plot(x, y)
title('$y = \sqrt[3]{x^7}$', 'interpreter', 'latex')
tmpStr = sprintf('$\\lambda = \\frac{\\phi}{%g}$', w);
ylabel(tmpStr, 'interpreter', 'latex')
text(0.6, 0.4, '$E[\xi]$', 'interpreter', 'latex')
```

Our solution to including the value of **w** (rather than the character "w") in the *y*-axis label is to use **sprintf** to produce a string containing the LaTeX expression we want. The string **tmpStr** will contain the string "$\lambda = \frac{\phi}{3.5}$". Note that we have to use "\\" in the **sprintf** command to produce a single backslash in **tmpStr**, because a single backslash is used to denote a special escaped character (such as "\n" for a newline character).

10.11   Arrows

Assume the vectors **xt** and **yt** contain (x, y) coordinates of the tails of some arrows, and vectors **xh** and **yh** contain the coordinates of the corresponding arrow heads.

R

The command arrows(xt, yt, xh, yh) will add the specified arrows to the figure. Appending the parameter **code = 3** will draw double-headed arrows instead.

MATLAB

The **annotation** function can be used, but note that it can only draw a single arrow at a time. If you have several arrows to add, you will need to use a loop to draw them all. For example:

```
————————————————— MATLAB —————————————————
for i = 1:length(xt)
  annotation('arrow', [xt(i) xh(i)], [yt(i) yh(i)])
end
```

There is another catch, however — the coordinates you provide must be in normalized figure coordinates, rather than in coordinates within the axes displayed in the figure. There are two solutions to this. First, you can search for the file implementing the function **dsxy2figxy** on The MathWorks File Exchange Web site; this will convert your (x, y) coordinates into the needed normalized figure coordinates. A second method is that you can instead set the parent of the arrow to the current figure axis. That is, this will do the trick:

```
                              MATLAB
for i = 1:length(xt)
  ah = annotation('arrow', [xt(i) xh(i)], [yt(i) yh(i)])
  set(ah, 'parent', gca)
end
```

Use **'doublearrow'** rather than **'arrow'** to produce double-headed arrows.

10.12   Figure legends

R

Producing a figure legend in R takes much more work than in MATLAB because you must do everything yourself. However, you also have greater control over what appears in the legend and where it is placed. The general form of the command is **legend(posStr, strVec)**, with one or more of several optional parameters also included. The **posStr** parameter is a position string indicating where to place the legend in the figure; possible values are **'topleft'**, **'top'**, **'topright'**, **'left'**, **'center'**, **'right'**, **'bottomleft'**, **'bottom'**, **'bottomright'**. The **strVec** is a vector of strings to display in the legend. The **legend** function interprets its parameters in a somewhat clever way, to allow you to specify x and y coordinates of either the top-left corner of the box, or two opposite corners. All coordinates are specified using the same coordinates axes or scales as the data plotted in the figure, rather than the 0–1 scale used by MATLAB for its legends. The way this works is that if a parameter named **legend** is not provided and the parameter named **y** does not contain numeric values, then the second parameter to the function is assumed to be the legend text (what I have called **strVec** above) and the first parameter has the coordinates. To place a legend with its top-left corner at coordinates (4.8, 15.16), you can use legend(4.8, 15.16, strVec). To place a legend with corners at (4.8, 15.16) and (23,42), you can either build a list with **x** and **y** elements via pos = list(x=c(4.8, 23), y=c(15.16, 42)) or a matrix with the **x** and **y** values in two columns via pos = matrix(c(4.8, 15.16, 23, 42), nrow=2, byrow=TRUE); you can then use legend(pos, posStr) to create the legend.

To specify what types of lines, plotting characters, colors and so on go into the legend, you can provide arguments **lty**, **pch**, and **col**. Each argument should be a vector giving the values to use for the various elements of the legend (values will be recycled if necessary; see page 23). Use an entry of **NA** if you do not want a line or plotting character for a particular entry. For example, assuming **strVec** has three strings, then legend(pos, strVec, lty=c(2, NA, 1), pch=c(NA, 2, 8)) produces a legend with three entries: the first entry has a dashed line (with no plotting character because of the **NA** in the first element of **pch**); the second entry has triangles (but no line because of the **NA** in the second element of **lty**), and the third entry has a solid line with stars.

There are many more parameters you can use with **legend**, to do things like add a title to the legend, specify the color of the text, control the arrangement of the items (for example horizontally or vertically), and so on.

MATLAB

MATLAB automates the production of figure legends for you much more than R does; this is usually more convenient, but also makes it a bit more challenging to craft a legend manually. Say you have plotted three things in the same figure window (and used `hold on` after the first plot, so they are all displayed together). You can then use the command `legend('first', 'second', 'third label')` to add a legend to the figure, showing the plotting markers and line styles for the three plots. By default, the legend is in the top-right corner. This can be changed with an additional string argument using compass directions (with **'NorthEast'** being the default top-right location). So for example, `legend('first', 'second', 'third label', 'Location', 'South')` will put the legend in the middle of the bottom of the plot. A location of **'Best'** will try to determine the best place inside the plot to put the legend.

You can also specify the location of the legend using a 1×4 position vector. This vector contains the (x, y) coordinates of the lower-left corner of the legend, along with the width and height. All four values in this vector range from 0 to 1 and are in terms of proportions within the figure window. For example, `legend('first', 'second', 'Location', [0.3 0.2 0.25 0.1])` places the legend 30% from the left and 20% from the bottom; the legend will have 25% of the width of the figure window and 10% of its height.

You may wish to hand-craft what appears in a legend. For example, you may only want the legend to display information about some of the things shown in the figure. In this case, you should save "handles" to your plots as you create them, as in the following sample code, and then pass those handles to **legend**:

```
—————————————— MATLAB ——————————————
h1 = plot(x1, y1, 'bo-')
hold on
h2 = plot(x2, y2, 'g-')
h3 = plot(x3, y3, 'r*')
legend([h1 h3], 'stuff', 'more stuff')
```

If for some reason you want the legend to display something which is not in the plot, you could plot something with the desired style, save the handle of that plot, and then make it invisible in the actual figure but pass the handle to **legend**:

```
—————————————— MATLAB ——————————————
h1 = plot(0, 0, 'bo-')
hold on
h2 = plot(0, 0, 'g-')
h3 = plot(x3, y3, 'r*')
h4 = plot(x4, y4, 'cs:')
set(h1, 'Visible', 'off')
set(h2, 'Visible', 'off')
legend([h1 h2], 'strange', 'more strange')
```

When you make the first two plots invisible, the axis automatically rescales if necessary, without the influence of the invisible points.

There are many more parameters you can use with **legend**, to do things like control the arrangement of the items (horizontally or vertically), turn the bounding box around the legend off, and so on.

10.13 Size and font adjustments

It is often useful to adjust the size of plotting characters or markers, the thickness of lines, the font size of axis labels, and so on. Both platforms give the user a fair amount of control over such things.

R

There are many parameters that can be passed to the **plot** (or **points** or **lines**) function to adjust the sizes of things. **cex=2** rescales the size of the plotting characters by a factor of 2. **lwd** specifies line width, but also the widths of lines used to draw plotting characters (circles, squares, etc.). If you want to use different line widths for the lines and the plotting characters, first use a command like **plot(x, y, type='c', lwd=3)** and then **points(x, y, lwd=2)** to draw them separately.

cex.axis rescales the numerical labels on the axes. **cex.lab** rescales the axis labels (those set with the **xlab** and **ylab** parameters). **cex.main** rescales the main plot title. **cex.sub** rescales the plot subtitle. For text labels added with the **text** function, use **cex=2** to double the font size.

For a figure legend, include **cex=2** to double the size of both the plotting characters and the legend text. You can use **pt.cex=3** to separately specify the size multiplier for the plotting characters and **title.cex** for the title text (if there is any) if you do not want them to share the same multiplier with the legend text.

For example, the following commands plot some data with customized sizes of pretty much everything in the plot, including different line thicknesses for the lines and the plotting characters:

```
─────────────────────── R ───────────────────────
plot(x, y, type='c', lwd=3, xlab='x values', ylab='dependent var',
  title='My data', sub='customized', cex.axis=2, cex.lab=3,
  cex.main=4, cex.sub=2, lwd=3)
points(x, y, cex=2)
```

MATLAB

Adjusting the sizes of things in MATLAB happens in a couple of different ways. Some properties are set via arguments to **plot**, whereas axis properties are adjusted by calls to **set**. Among arguments to **plot**, adding the pair of parameters **'MarkerSize', 10** sets the size of the plotting marker. Including the parameters **'LineWidth', 2** changes the line width. This affects both the line width, and the width of lines used to draw the plotting markers. To use different line widths for lines and plotting markers, first plot only the lines, followed by **hold on**, and then add the markers.

Use set(gca, 'FontSize', 20) to adjust the font size of numerical labels on the axes. (gca stands for "get current axis," and returns a handle to the axis; use just get(gca) to see a list of properties you can **set**.) xlabel('xfoo', 'FontSize', 20) specifies the x-axis label font size; and similarly with **ylabel** and **title**. For added text labels, use text(x, y, 'some text', 'FontSize', 20) to specify font size. Alternatively, when you create the axis labels or added text, you can save handles for them, and set things via the handles. For example, h1 = xlabel('xfoo'); set(h1, 'FontSize', 25) and similarly with h2 = text(3, 4, 'hmmmm'); set(h2, 'FontSize', 18). You must use the handle approach to adjust font size of legend text: h3 = legend('foo', 'bar'); set(h3, 'FontSize', 25).

10.14   Two y axes

Sometimes, when plotting two data sets containing y values of fairly different magnitudes, it is useful to have two separate y axes in a plot. One axis is drawn on the left side of the figure, and another on the right side. Here we will plot the functions $y_1 = \sin(x)$ as x varies between 0 and 10, and $y_2 = 5e^{-x}\cos(10x)$ as x varies between 0.5 and 8. This takes somewhat complicated manual fiddling with things like axes and margin text in R to accomplish; in MATLAB, the **plotyy** function does the hard work. The resulting plots are shown in Figure 10.2.

R

```
R
x1=seq(0,10,len=200)
y1=sin(x1)
x2=seq(0.5,8,len=200)
y2=5*exp(-x2)*cos(10*x2)
par(mar=c(5.1, 4.1, 4.1, 5.1))
plot(x1, y1, type='l')
par(new=TRUE)
plot(x2, y2, type='l', lty=2, xaxt='n', yaxt='n', xlab='', ylab='')
axis(4)
mtext('y2', side=4, line=3)
legend('top', legend=c('y1', 'y2'), lty=c(1,2))
```

MATLAB

```
MATLAB
x1=linspace(0,10,200);
y1=sin(x1);
x2=linspace(0.5,8,200);
y2=5*exp(-x2).*cos(10*x2);
[ax,h1,h2]=plotyy(x1,y1,x2,y2);
set(h1, 'color', [0 0 0])  % black line
set(h2, 'color', [0 0 0], 'linestyle', '--')  % black dashed line
set(ax(1), 'ycolor', [0 0 0])  % black y-axis1 labels
set(ax(2), 'ycolor', [0 0 0])  % black y-axis2 labels
legend('y1', 'y2', 'location', 'north')
```

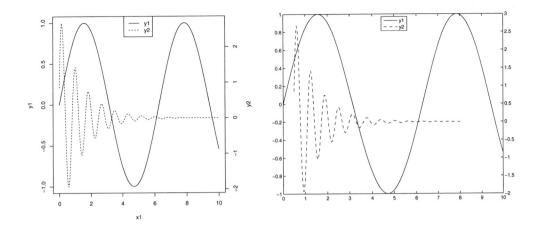

FIGURE 10.2
Figures with two y axes, produced with the R (left) and MATLAB (right) commands from
Section 10.14.

10.15 Plotting functions

Sometimes you may wish to plot a function $y = f(x)$ over some range of values of x. A simple
way to do so is to generate a vector **x** containing some values of x (with equally spaced
values being the straightforward way), compute a vector **y** containing the corresponding
function values, and then plot the values stored in **x** and **y**. Choosing how many values you
need in **x** to produce a nice-looking plot may take some trial and error. R does not have a
built-in function to aid you in doing this. It does have a function **curve** that you can use to
plot a function at equally spaced points, without the need to explicitly set up your own **x**
and **y** vectors. For example, `curve(sin, 0, 15)` plots the function $\sin(x)$ with x ranging
from 0 to 15 (with 101 points by default, but you can use the parameter **n** to modify that).

MATLAB provides a nice function **fplot** to perform this task somewhat automatically.
You provide a function and the minimum and maximum values of x, and it determines
how many values of x to use, along with their spacing. It places the x values more closely
together where the function is varying more rapidly. You can either specify the function to
plot as a string, or via a function handle. For example, to plot $\sin(x)$ from 0 to 4π, you can
either use `fplot(@sin, [0 4*pi])` or `fplot('sin(x)', [0 4*pi])`. To plot the function
$\sin(1/x)/x$ from 0.01 to 0.2 with red points and lines, you can use one of the following two
commands:

```
———————————————— MATLAB ————————————————
fplot(@(x) sin(1/x)/x, [0.01 0.2], 'r.-')
fplot('sin(1/x)/x', [0.01 0.2], 'r.-')
```

The first command above uses an anonymous function (see Section 8.2.1). **fplot** also lets
you specify a minimum number of points to use (the default minimum is 2) and a relative
error tolerance which adjusts how many points are used.

10.16   Image plots and contours

There are many other types of plots besides the basic scatterplots or line plots already discussed. There are 3-D surface plots, pie charts, and many others. One common type of plot is an image plot of some data in a matrix. There are two ways image plots are typically made: the matrix contains discrete integer values that are used as indices into a colormap, or the matrix contains continuous values that are plotted as different colors in a colormap. The x and y coordinates corresponding to the matrix entries may simply be the integers $1, 2, 3, \ldots$, or they may be continuous values from within some range.

To have some concrete examples to work with, we will set up three matrices. Let $\mathbf{A}$ be a 10×10 matrix containing the values 1 through 10 arranged so that the value in row i, column j is simply j, but we will then change two entries in the four corners to have the values 1, 2, 3, and 4 (counterclockwise from the top left, i.e., $a_{1,1} = a_{2,2} = 1$, $a_{10,1} = a_{9,2} = 2$, $a_{10,10} = a_{9,9} = 3$, and $a_{1,10} = a_{2,9} = 4$) to better see the orientation of how $\mathbf{A}$ is plotted. Next, let $\mathbf{A2}$ be the same as $\mathbf{A}$, but all values greater than 5 will be replaced by the value 5. And finally, $\mathbf{B}$ will be a matrix containing values of the function $z = f(x, y) = \cos(xy)\sin(x^2 + y^3)$, which will be computed for 200 values of x equally spaced between 1 and 4, and 200 values of y between -3 and +3.

R

The following commands will set up the three matrices we will be using, along with two rainbow colormaps. The first colormap has 10 entries and the second has 250 entries. The **end=5/6** parameter to **rainbow** tells it to begin with red and end with violet, rather than beginning and ending with red. Two additional colormaps are also generated, containing shades of gray.

```
R
A = matrix(rep(1:10, 10), nrow=10, byrow=TRUE)
A[1,1] = 1; A[2,2] = 1;   A[10,1] = 2; A[9,2] = 2;
A[10,10] = 3; A[9,9] = 3; A[1,10] = 4; A[2,9] = 4
A2 = A;   A2[A2 > 5] = 5
x=seq(1, 4, len=200)
y=seq(-3, 3, len=200)
B=cos(outer(x, y)) * sin(outer(x^2, y^3, '+'))
cmap1 = rainbow(10, end=5/6);   cmap2 = rainbow(250, end=5/6)
cmap1g = gray(seq(0, 1, len=10));   cmap2g = gray(seq(0, 1, len=250))
```

You can use the command `image(A)` to plot an image of $\mathbf{A}$, or `image(A, col=cmap1)` to plot it using the small rainbow colormap (the default colormap is the one produced by **heat.colors(12)**). Doing so and examining the colors in the four corners shows that the image is displayed rotated 90° from the way the values are displayed if you simply type A at the R command prompt. That is, column 1 of the matrix is plotted along the bottom row of the image. The (row, column) indices in the matrix get used as (x, y) coordinates in the image, after rescaling them to lie between 0 and 1. A grayscale version of the figure is shown in Figure 10.3.

If you wish to display the matrix in the orientation used by MATLAB's **image** command, you need to reverse the direction of the y axis (which can be done by specifying the **ylim** parameter with the larger value followed by the smaller one) as well as transpose the matrix. The **image** function lets the x and y coordinates range from 0 to 1 by default. The cells are

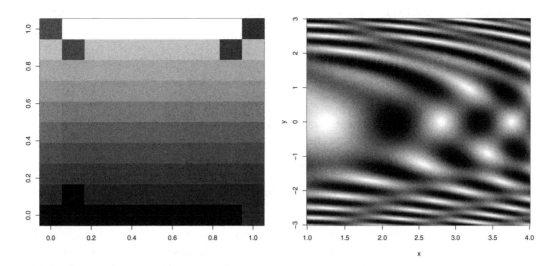

FIGURE 10.3
Left: An image plot produced in R via `image(A, col=cmap1g)`. Right: An image plot produced via `image(B, col=cmap2g)`.

centered at values in that range, with widths of $1/m$ and heights of $1/n$ for an $m \times n$ matrix. That means when you transpose the matrix the y coordinates will range from $-1/(2n)$ to $1+1/(2n)$, so the following command will display the matrix **A** in MATLAB's orientation: `m=dim(A)[1]; image(t(A), col=cmap1, ylim=c(1+1/(2*m), -1/(2*m)))`.

If you look at the output produced by `image(A2, col=cmap1)`, you will notice that the colors still range over the full red to violet spread of the colormap, even though the values only range from 1 to 5 (rather than 1 to 9 as in the matrix **A**). By default, **image** rescales and maps the values in the matrix being plotted onto the entire range of the colormap. You can instead explicitly specify "break points," i.e., cutoff values indicating which ranges of values should be displayed using each color. If your colormap has n entries, you must specify $n+1$ break points. If the break points are in vector **b**, then the ith color in the colormap is used to display values in the matrix that lie between **b**[i] and **b**[i+**1**]. The first interval (**b**[**1**] to **b**[**2**]) includes both of its boundaries; subsequent intervals contain only their upper boundaries. Using the vector **0:10** for the break points will display the value 1 using the first color (all values from 0 to 1 inclusive would be drawn using this color), the value 2 using the second color, and so on. Any values which do not fall within the range of values specified by break points are not drawn. The command `image(A2, col=cmap1, breaks=0:10)` therefore plots the data using the values in **A2** as indices into the colormap.

To plot the $z = f(x, y)$ values in the matrix **B**, with x and y coordinates taken from the vectors **x** and **y**, the command `image(x, y, B, col=cmap2)` can be used. A grayscale version is shown in Figure 10.3.

One limitation of graphics produced by the **image** command is that they do not contain a legend indicating which colors represent which values (as you can produce in MATLAB via the `colorbar` command). The **levelplot** function in the **lattice** package will produce image plots with a colorbar (see Section 13.9 for how to load a package; the **lattice** package is included in default installations of R, but not loaded). To reproduce the plot in the right part of Figure 10.3 but with a colorbar legend added, a command like this is needed: `levelplot(B, row.values=x, column.values=y, aspect='fill',`

xlim=c(min(x), max(x)), ylim=c(min(y), max(y)), col.regions=cmapg, cuts=249).
The parameters are as follows:

row.values, column.values: Vectors containing the x and y values, respectively.

aspect: We specify **'fill'** so the plot fills the figure window; the default when using **levelplot** with a matrix is to make each element of the image square.

xlim,ylim: If the x and y limits are not specified, the plotted image will not fill the figure. The detailed appearance of the results will depend on the range of x and y coordinates.

col.regions: Specifies the colormap to use.

cuts: How many breaks there are between colors. With 250 entries in our colormap, there are 249 breaks between the colors.

Instead of **cuts** to request equally spaced cutoffs in z values for the colors, you can specify **at** to give an explicit vector of cutoffs. This works like the **breaks** argument to **image**, so if you want 250 levels, your **at** vector should have 251 values, e.g., **at=seq(-1, 1, len=251)**.
    You can also provide a **colorkey** parameter to **levelplot**, containing a list with detailed information about the color legend. This can be used to do things such as construct a color scale with logarithmically spaced intervals for the colors.

MATLAB

The following commands set up the three matrices we will be using, along with two rainbow colormaps. The first colormap has 10 entries and the second has 250 entries. The command **cmap = hsv(10)** would set up a rainbow colormap with 10 entries which begins and ends with red. To generate one that begins with red and ends with violet, here we first build a 3-column HSV (hue-saturation-value) matrix where the hue goes from 0 to 5/6 with the saturation and value set to 1, and convert it to RGB (red-green-blue). An alternative would be to make a colormap 6/5 times as long as we want, and then take the first 5/6 entries to get a red-violet rainbow, using the commands **tmp=hsv(300); cmap2=tmp(1:250,:);**.

```
─────────────────────────── MATLAB ───────────────────────────
A = repmat(1:10, 10, 1);
A(1,1) = 1; A(2,2) = 1; A(10,1) = 2; A(9,2) = 2;
A(10,10) = 3; A(9,9) = 3; A(1,10) = 4; A(2,9) = 4;
A2= A;  A2(A2 > 5) = 5;
B = repmat(1:100, 100, 1);
x = linspace(1, 4, 200);
y = linspace(-3, 3, 200);
[X,Y] = meshgrid(x,y);
B = cos(X .* Y) .* sin(X .^ 2 + Y .^ 3);
cmap1 = hsv2rgb([linspace(0, 5/6, 10)' ones(10, 2)]);
cmap2 = hsv2rgb([linspace(0, 5/6, 250)' ones(250, 2)]);
cmap1g = gray(10);  cmap2g = gray(250);
```

You can use **image(A)** to plot an image of **A** with the default colormap (the one produced by **jet(64)**). After making the plot, the command **colormap(cmap1)** can be used to switch to the small rainbow colormap. Doing so and examining the colors in the four corners shows that the image displays the values in the same arrangement that you would see by simply typing **A** at the MATLAB command prompt. That is, the values in the last (bottom) row of the matrix go across the bottom of the image. The results are shown in Figure 10.4.

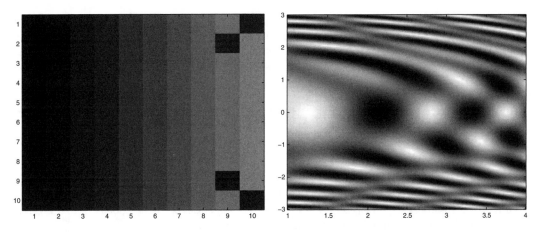

FIGURE 10.4

Left: An image plot produced in MATLAB via `image(A); colormap(cmap1g)`. Right: An image plot produced via `imagesc(x, y, B); colormap(map2g); set(gca, 'YDir', 'normal')`.

Also note that the y axis label increases from top to bottom (as the rows of a matrix are numbered), rather than from bottom to top as usual. This is because **image** sets the **YDir** property of the axis to **'reverse'**. You can flip it upside down via the command `set(gca, 'YDir', 'normal')`. To plot the matrix in the same orientation as R's **image** function, instead plot the transpose of **A** and then flip the **YDir**, i.e., use `image(A'); set(gca, 'YDir', 'normal')`.

The command `colorbar` adds a very useful color legend to the right side of the figure. The **colorbar** function can take optional arguments to specify the location or other properties of the legend, for example, `colorbar('FontSize', 14)`.

MATLAB does not have a simple method for specifying break points to use with the colors when display a matrix using the **image** function. Instead, the values in the array are simply used as indices into the current colormap. (For values of type **double**, the indices should be in the range $1, 2, \ldots$; for values of type **uint8** or **uint16**, they should be in the range $0, 1, \ldots$.) Instead, you can build a vector of break points, and then use **histc** to build a matrix of values which can be used as indices into the colormap. For example, if you want three bins to represent values satisfying $0 \le x < 2$, $2 \le x < 3$, and $3 \le x \le 7$, then use the following commands:

```MATLAB
edges = [0 2 3 7];
[tmp, binidx] = histc(A, edges);
image(binidx)
```

If you have continuous values in your matrix, the **imagesc** function will automatically rescale your values so that the full range of the colormap corresponds to the range of values in the matrix. The **pcolor** function works like **imagesc**, but the **YDir** attribute is set to **normal** by default, and the last row and column of the matrix are not drawn. **pcolor** also has a number of shading mechanisms available. The default is **faceted**, in which the last row and column are not drawn, and there are thin black lines drawn between the cells of the matrix. The command `shading flat` will switch to flat shading, which eliminates the black lines. `shading interp` uses linear interpolation within each cell, according to the values at the cell's corners.

To plot the $z = f(x, y)$ values in the matrix **B** set up earlier, with x and y coordinates taken from the vectors **x** and **y**, the commands `imagesc(x, y, B); colormap(cmap2);` `set(gca, 'YDir', 'normal')` can be used. A grayscale version is shown in Figure 10.4.

10.17   Colormaps

In R, built-in functions to provide various colormaps are **rainbow**, **heat.colors**, **terrain.colors**, **topo.colors**, **cm.colors**, and **gray**. You can also build a colormap yourself. For example, a colormap containing the colors red, magenta, tan, and green could be built via `mycmap = c('#ff0000', '#ff00ff', '#d2b48c', '#00ff00')`. The functions **colors**, **palette**, **hsv**, **hcl**, **rgb**, **col2rgb**, and **convertColor** may also be useful, for example for converting a set of color values encoded as numerical values from 0 to 1 into the above format. The above colormap could be built by first constructing a 4×3 matrix, with each row containing the intensities of red, green, and blue (ranging from 0 to 1), and then converting it as follows:

```R
A=matrix(c(1,0,0,1,0,1,210/255,180/255,140/255,0,1,0),nrow=4,byrow=TRUE)
mycmap = rgb(A[,1], A[,2], A[,3])
```

In MATLAB, built-in colormaps are provided by the functions **jet**, **hsv**, **hot**, **cool**, **pink**, **flag**, **spring**, **summer**, **autumn**, **winter**, **gray**, **bone**, **copper**, and **lines**. A colormap containing the colors red, magenta, tan, and green could be built by constructing a 4×3 matrix, with each row giving intensities of red, green, and blue: `mycmap=[1 0 0 ; 1 0 1 ; 210/255 180/255 140/255; 0 1 0]`. The functions **rgbplot**, **hsv2rgb**, and **rgb2gray** may also be useful.

10.18   3-D plotting

Three-dimensional surface plots are available in both packages; these can be used to plot a function of two variables, $z = f(x, y)$. MATLAB colors the surface according to the height z by default; R does not do this automatically, but you can provide a color matrix to duplicate this functionality. Here we will plot the function $z = e^{-x} \sin(3y)$ for 30 equally spaced x values between 1 and 4, and 80 equally spaced y values between -3 and +3.

R

There is a function **persp** in the traditional graphics system which produces 3-D surface plots. You can provide vectors **x** and **y** containing the independent variables and a matrix **z**, along with usual parameters like **xlab** to label the x axis, **ylim** to specify the range of values shown on the y axis, and so on. By default, the axes simply have arrows showing the direction in which they increase, without numerical labels. You can use **ticktype='detailed'** to get back the standard numerical labels. The parameters **theta** and **phi** can be used to specify the viewing angle. Other parameters are available, including **shade** to apply shading to the surface.

If you provide an $m_x \times m_y$ matrix (as with the **image** function, the rows correspond to x

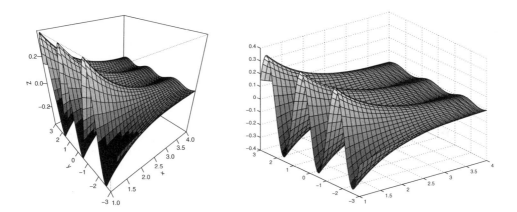

FIGURE 10.5
Surface plots of the function $z = e^{-x}\sin(3y)$. Left: in R using **persp**. Right: in MATLAB using **surf**.

coordinates and the columns correspond to y), an $(m_x - 1) \times (m_y - 1)$ mesh of surface facets is produced. If you wish to provide a colormap matrix, it should therefore be of that size. To color the facets according to the heights z, you can first use either **cut** or **findInterval** to transform the z matrix into a set of values indicating which bin each z value lies within. That matrix of bin numbers can then be used to index a colormap, producing a colormap matrix. Commands to do this for our example function above are below. The **outer** function is used to build the matrix **Z** from the vectors **x** and **y**.

```
————————————————————— R —————————————————————
x = seq(1, 4, len=30)
y = seq(-3, 3, len=80)
Z=outer(exp(-x), sin(3*y))
cmapLength = 20
# reverse violet-to-red rainbow colormap similar to Matlab "jet" colormap
cmap = rev(rainbow(cmapLength, end=5/6))
cmapg = gray(seq(0, 1, len=cmapLength))  # or use gray colormap
# Calculate matrix of bins for Z values
edges = seq(min(Z), max(Z), len=cmapLength)  # edges of the bins
bins=array(findInterval(Z, edges, rightmost.closed=TRUE), dim=dim(Z))
bins = bins[1:(length(x)-1),1:(length(y)-1)]
persp(x, y, Z, col=cmap[bins], theta=-40, phi=30, ticktype='detailed')
```

This produces the image in the left half of Figure 10.5 when used with the **cmapg** gray colormap.

The **lattice** package has a function **wireframe** that can be used to produce a similar plot. One should first build a data frame containing all pairs of (x, y) coordinates, and then create a **z** column. This can be done via the following commands, assuming **x** and **y** have been set up as above:

```
───────────────────────────── R ─────────────────────────────
g=expand.grid(x=x,y=y)
g$z=exp(-g$x)*sin(3*g$y)
> wireframe(z~x*y, g)
```

MATLAB

In MATLAB, the **surf** function will produce a surface plot. You provide a matrix **Z** containing the z coordinates, whose rows correspond to the y coordinates and columns correspond to x (this is the opposite of R). It is convenient to use **meshgrid** to build matrices **X** and **Y** containing the x and y coordinates, and then use element-by-element matrix calculations to produce **Z**, as shown below.

```
───────────────────────────── MATLAB ─────────────────────────────
x = linspace(1, 4, 30);
y = linspace(-3, 3, 80);
[X,Y] = meshgrid(x, y);
Z = exp(-X) .* sin(3*Y);
surf(X,Y,Z)
% Then use, e.g., colormap(jet) or colormap(gray)
```

This produces the image in the right half of Figure 10.5.

For variations on this type of plot, see the **mesh** and **waterfall** functions. You can also use **surfc** to produce a surface plot with contour lines, or **surfl** to produce a shaded surface with lighting.

10.19   Multiple subplots in one figure

It is often desirable to have several smaller graphs within a single figure. Depending on the details, I typically save the different graphs in separate files and then combine the images when formatting my document. For example, the two images in Figure 10.1 were placed beside each other using a LaTeX **tabular** environment. But sometimes it is more convenient to have several smaller graphs within a single figure window in the software itself. MATLAB makes this relatively easy to do. R provides a couple of ways to do it which may not be quite as simple as MATLAB's methods, but which give more flexibility.

R

There are three approaches to using subfigures in the traditional graphics system. The three approaches are incompatible with one another; you cannot mix and match.

The first approach is to use the command `par(mfrow=c(m, n))` to request an $m \times n$ grid of equally sized subfigures (m rows, n columns). Subsequent **plot** commands will draw things in consecutive subfigures; **lines** and **points** can be used as usual to add things to an existing subfigure. For example, `par(mfrow=c(1,2)); plot(x1, y1); lines(x2, y2); plot(x3, y3)` will plot the first two data sets in the first subfigure, and the third data set in the second subfigure.

The second approach is to use the **layout** function. This is most similar to MATLAB's mechanism for working with subfigures. You first create a layout matrix describing the arrangement of the subfigures. For example, consider the matrix

$$A = \begin{bmatrix} 1 & 1 & 2 \\ 1 & 1 & 3 \\ 4 & 5 & 5 \end{bmatrix}.$$

The matrix is 3×3, so the figure window will be conceptually divided into a 3×3 array of cells. The values in the matrix A show that the upper-left 2×2 block of cells will correspond to subfigure 1, the lower-right 1×2 block will correspond to subfigure 5, and so on. The matrix is a visual map showing which subfigures are in which locations. Consecutive **plot** commands will draw in consecutive subfigures, as when using **par(mfrow)**. The following sequence of commands will produce plots of some random data in this layout, as seen in Figure 10.6.

—————— R ——————

```
A = matrix(c(1,1,2,1,1,3,4,5,5), nrow=3, byrow=TRUE)
layout(A)
plot(runif(25))
lines(runif(25))
plot(runif(10))
plot(runif(15), pch=4)
plot(runif(50), pch=6)
plot(runif(20), type='b')
```

Note that you can put the value 0 in the matrix **A** to mark locations where you do not want any subfigures, if you want to leave some "holes" in the final figure.

To reset the figure window back to using a single subfigure, you can use the command `par(mfrow=c(1,1))` to request one row and one column.

The third approach to subfigures in R's traditional graphics system is with the **split.screen** function and some related functions. You can use **split.screen** to divide the graphics device into subfigures; you can then further divide some of those into smaller subfigures. You can either pass in a vector containing the number of rows and columns to divide things into, or pass in a 4-column matrix where each row contains (left, right, bottom, top) values (in the interval from 0 at the bottom/left to 1 at the top/right) describing one of the subfigures. For example, in the image shown in Figure 10.6, there are five subfigures with the following ranges:

Subfigure	Left	Right	Bottom	Top
Top-left large	0	2/3	1/3	1
Top-right small	2/3	1	2/3	1
Middle-right small	2/3	1	1/3	2/3
Bottom-left small	0	1/3	0	1/3
Bottom-right wide	1/3	1	0	1/3

Calling `v = split.screen(A)` with the above 5×4 matrix would divide the figure window as in Figure 10.6. The return value stored in the vector **v** contains the screen numbers. You can then use `screen(n)` to select screen number **n** for drawing in, `erase.screen(n)` to erase a screen, or `close.screen(all.screens = TRUE)` to close all screens and go back to regular plotting. See the warnings in the help for **split.screens**, including the fact that you should not draw something in one screen (subfigure), draw something in a different screen, and then go back and try to add more things to the first screen.

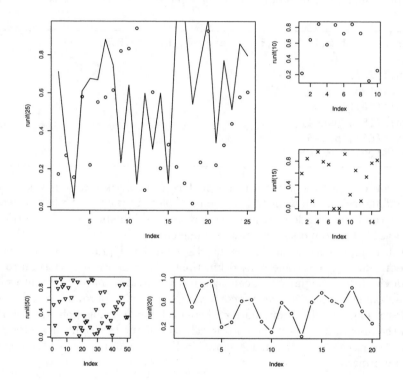

FIGURE 10.6
Subfigures produced with the **layout** function in R.

As an example of further subdividing screens, you can create a 2×1 grid of screens, then subdivide the second one into a 1×2 grid of smaller screens, via the following commands:

```R
v1 = screen(c(2,1))   # v1 contains the values 1 and 2
v2 = screen(c(1,2), 2)  # v2 contains the values 3 and 4
screen(1); plot(runif(5))    # plot in screen 1
screen(3); plot(runif(15))  # plot in screen 3
screen(4); plot(runif(30))  # plot in screen 4
```

Note that after doing the above commands, you could then still use the command `screen(2)` and plot things in screen 2, even though it occupies the same locations as screens 3 and 4. The **screen** mechanism will not stop you from doing strange things like this.

MATLAB

The **subplot** function is used to divide a figure window into smaller portions. One way to call it is as `subplot(m, n, p)`. This divides the figure window into an $m \times n$ array of subfigures, and selects the pth one to draw in. The subfigures are counted across the top row, then across the second row, and so on. `p` can be a vector containing several such positions, in which case the subfigure will span all of them. Below is an example, with the results shown in Figure 10.7.

```MATLAB
clf    % clear figure window
subplot(3,3,[1 2 4 5])
plot(rand(25,1),'o')
hold on
plot(rand(25,1),'-')
subplot(3,3,3)
plot(rand(10,1),'o')
subplot(3,3,6)
plot(rand(15,1),'x')
subplot(3,3,7)
plot(rand(50,1),'v')
subplot(3,3,[8 9])
plot(rand(20,1),'o-')
```

An alternative way to use **subplot** is to instead call `subplot('position', [left bottom width height])`, where the four values in the position vector are given in normalized coordinates which range from 0 to 1.

You can use the command `clf` to clear a figure window at any time, which among other things resets the subplot structure of the window so that the next plotting command uses the entire figure window.

10.20   Saving figures

This section describes how to save the contents of a figure to a file, in various formats. Of course, you may also be able to save the contents of a figure window via the user-interface, and you may find that more convenient for casual use. If the figure window is active, either

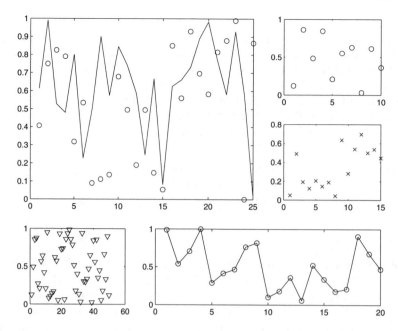

FIGURE 10.7
Subfigures produced with the **subplot** function in MATLAB.

right-clicking in the figure, or going to the File menu may provide various options for saving. But as usual, the commands described in this section provide for a way which can be automated in your program scripts. In my own work, it is not usual to produce a figure in PDF form for a journal article, and then several months later be asked by one of the reviewers to make some changes to the figure such as changing the font size. I typically write a separate script to produce each figure and save the output in PDF form, so that I can simply edit the appropriate script to make the requested changes to the figure.

R

There are two primary ways to save a figure to a file. You can either produce a figure using one of the graphics devices (e.g., Windows, Quartz, or X11) available under your operating system and then copy the contents of the figure window to a file, or else you can directly open a file (using a format such as PDF, JPG, etc.) and then "draw" directly to the file. If you are working interactively at the R console, the first method is most convenient, as you can see what you are doing while building the plot before saving the results. If you are producing your graphics via a script, either method is convenient. Note that fonts and some details may differ between the two methods of saving graphics, so if you are having trouble with one method, you may wish to try the other.

To copy the contents of the active figure window to a file, you can use the **dev.copy** function. The first argument should be a device. The second argument can be a filename; if none is provided, **RplotXXX** will be used (with an appropriate suffix), where XXX is the first three-digit number for which a saved file does not already exist. For example, dev.copy(jpeg, 'foo.jpg') saves the active figure to the file **foo.jpg** in JPEG format. The devices **pdf** and **postscript** are generally most useful for those working with LaTeX. If using **postscript**, you may wish to include the parameter **horizontal = FALSE** to force it to use portrait mode rather than landscape mode. The **win.metafile** device (only available under Windows) produces a WMF file which may be useful for those working with Microsoft Office. Devices **jpeg**, **png**, **bmp**, and **tiff** are useful for those who prefer to work with raster images. When using **jpeg**, you can also specify a **quality** parameter (as a percentage, e.g., **quality=90**), with higher quality values producing larger files with better-quality images. After using **dev.copy** to save an image, you must then call `dev.off()` to turn off the new graphics device that gets created during the copying, in order to flush the data to the file. Also see the files **dev.print**, **dev.copy2eps**, and **dev.copy2pdf**, including information about sizing the resulting saved images.

Also note, some people may use `dev.control(displaylist = 'inhibit')` to disable an internal display list which is used to redraw the contents of a figure window if the window is resized; disabling this feature reduces memory usage. However, if this has been done, **dev.copy** will not function properly, because it uses the display list.

The second method of saving graphics to a file is to simply call a function named according to one of the above devices, use all of your **plot** or other commands to produce the graphics, and then call **dev.off()** to flush the output and close the file. For example, `pdf('foo1.pdf'); plot(runif(50)); dev.off()`. Or `jpeg('foo1.jpg', quality=85); plot(runif(50)); dev.off()`.

MATLAB

The **print** function allows you to save the contents of the current (or a specified) figure window to a file. For example, `print -deps foo1.pdf` saves to the file **foo1.pdf** using the Encapsulated PostScript device driver. Other drivers include:

- **-depsc**: Encapsulated Color PostScript

- **-dpdf**: PDF

- **-dmeta**: Windows Metafile format (available under Windows only)

- **-djpegXX**: JPEG with specified quality level, e.g., **-djpeg90** for a quality level of 90

- **-dpng**: PNG

- **-dtiff**: TIFF

10.21   Other types of plots

There are of course a huge variety of other types of plots one may wish to make.

Bar plots: Assume you have vectors **x** and **y**, respectively, containing the x coordinates and the heights of some bars you wish to draw. In R, you can use `plot(x, y, type='h', lwd=10)` to draw a histogram-style plot with lines made thick enough to look like bars. In MATLAB, you can simply use `bar(x, y)`.

Histograms: In both platforms, `hist(v)` will produce a histogram of values in the vector v. To produce a histogram with k bins, you can try `hist(v, k)`.

However, note that in R, the number of bins **k** you specify is treated as merely a "suggestion," and it often does not actually produce the number of bins you request. You can force it to use the desired number of bins by instead giving a "breaks" vector (say, **b**) containing the breakpoints between the bins. Because the ith bin goes from **b[i]** to **b[i+1]**, to obtain k bins your breaks vector should be of length $k + 1$. So for example, `hist(v, seq(min(x), max(x), len=k+1))` will produce a histogram with **k** bins.

Pie charts: `pie(v)` will produce a pie chart using the values in vector **v** in both platforms.

Filled circles of various sizes and colors: Say **x** and **y** are vectors containing coordinate values, **s** contains values between 1 and 10, and **c** contains values from 1 to 8. In R, `plot(x, y, pch=21, cex=s, col=c, bg=c)` produces this graph, using **s** to specify the sizes and **c** to specify the colors. See the **palette** function (i.e., call `palette()`) to see which colors are referred to by the color indices. One could instead build a vector **c** of color strings (like `'#FFFF00'` for yellow). Note that plotting characters 21–25 can be filled in this manner; the **col** parameter gives the color of the border, while **bg** gives the internal color (gray by default). An alternative way to do this in R is via `symbols(x, y, circles=s, bg=c, fg=c)`. The **symbols** function can produce other shapes as well.

In MATLAB, `scatter(x, y, s, c, 'filled')` will produce the desired plot.

Error bars: While you could of course construct a plot containing error bars yourself by drawing the various line segments, both platforms provide routines to facilitate this. Let us say we wish to plot a curve with (x, y) coordinates stored in vectors **x** and **y**, while vectors **a** and **b** give the desired corresponding heights of the error bars above and below that curve, respectively. Note below that R and MATLAB handle their arguments specifying the error bars slightly differently, and in opposite order for the heights above/below the main curve.

In R, you can use `errbar(x, y, y+a, y-b)`; this function is part of the **Hmisc** package (this package is not included in the standard R installation; see Section 13.9 for more about installing and loading packages). In MATLAB, you can use `errorbar(x, y, b, a)`. If you want the error bars to have the same height **s** above and below the curve in MATLAB, just `errorbar(x, y, s)` will work. You can specify the usual plotting style for the main curve itself, e.g., `errorbar(x, y, b, a, 'o--')` to use circles and a dashed line.

For additional control over plotting styles, e.g., to use different line thicknesses for the main curve and the error bars, you may wish to construct the plot in two separate pieces for the main curve and the error bars. In MATLAB, you can of course use **hold on** to do this. In R, including the parameter **add=TRUE** in the call to **errbar** will add a new error bar to the existing plot, rather than creating a new plot.

3-D plots: In R, the traditional graphics system does not provide any nice 3-D plotting routines. Some alternatives are: `scatterplot3d(x, y, z)` from the **scatterplot3d** package, `cloud(z~x*y)` from the **lattice** package, and `plot3d(x, y, z)` from the **rgl** package. The **rgl** package also lets you interactively rotate the 3-D plot using your mouse. In MATLAB, you can use `plot3(x, y, z)`. By clicking on the small icon showing a circular arrow around a cube, you can then rotate the plot using your mouse.

Contour plots: Both platforms have a **contour** function which works much like the surface plotting functions **persp** (in R) and **surf** (in MATLAB), in terms of how to set up the data and call the function `contour(x, y, Z)`.

Other: The MATLAB command `stairs(x, y)` produces a figure equivalent to R's `plot(x, y, type='s')`. The MATLAB command `stem(x, y)` produces a plot equivalent to the R commands `plot(x, y, type='h')`; `points(x, y)`.

10.22 Final notes about graphics

R

As pointed out in Section 7.3, some graphics functions, particularly those in the **lattice** package (such as **levelplot**) do not display their results when called from within scripts, functions, or loops such as **for** loops. To ensure they are displayed in those situations, rather than simply calling **levelplot(...)**, you should instead use **print(levelplot(...))**.

MATLAB

MATLAB typically only updates figure windows when control returns to the MATLAB command window after a script or function finishes, along with a few other circumstances. Usually, this is not a problem. However, if you write a script or function which performs some iterative calculations and draws some graphics along the way, you may wish to see the graphics as the code is running. In that case, you can call the command `drawnow`, which flushes the event queue and updates figures. (See the help for **drawnow** for the other cases in which the figure windows will be updated.)

11

Numerical Computing

Both platforms contain facilities for performing standard numeric computations such as root-finding, function optimization, numerical integration or quadrature, numerically solving differential equations, and so on. MATLAB has a longer history of use in scientific computing, and there are a number of standard numerical analysis textbooks whose examples are all based around MATLAB [4, 7, 10, 17, 18, 19, 27]. The textbook market in this area has not yet matured for R, though reference [14] includes some material on numerical computing.

11.1 Root-finding

Given a function of one variable, $f(x)$, one may want to find a value x satisfying $f(x) = 0$. Both platforms provide straightforward ways to do this.

One special case is when the function f is a polynomial of degree n, $f(x) = a_n x^n + a_{n-1} x^{n-1} + \ldots + a_2 x^2 + a_1 x + a_0$. Say the vector **vecd** contains the coefficients $a_n, a_{n-1}, \ldots, a_1, a_0$, in descending order, and the vector **veci** contains the coefficients $a_0, a_1, \ldots, a_{n-1}, a_n$ in increasing order. In R, you can find the roots of the polynomial using `polyroot(veci)`, while in MATLAB, use `roots(vecd)`. See entry 22 in Section 3.3 if you need to reverse the order of the coefficients in your vector.

To find a zero of a general function $f(x)$, first define an R or MATLAB function of a single variable, as in Section 8.1.1 or Section 8.2.2. Suppose this function is named **myfunc** and that you want to search for the zero among all values of x satisfying $a \le x \le b$.

In R, use `uniroot(myfunc, c(a,b))`. This returns a list containing the estimated root $\hat{x}$, the value of the function $f(\hat{x})$ at that point, the number of iterations used, and an estimate of the precision of the root. You can extract the root via `uniroot(myfunc, c(a,b))$root` or the function value there via `uniroot(myfunc, c(a,b))$f.root`.

In MATLAB, use `fzero(@myfunc, [a b])`. (The notation **@myfunc** builds a "function handle" to pass your function to **fzero**, rather than trying to call **myfunc** directly. See Section 8.2.7 for more information about function handles.) This returns only the estimate of the root. You can use multiple return values to receive extra information. For example, `[x,fval,exitflag,output]=fzero(@myfunc,[a b])` returns an estimate **x** of the root, the function value $f(\hat{x})$ there, an exit flag (which will be 1 if it found a root), and an output structure indicating how many iterations and how many calls to your function were used. MATLAB also lets you provide just a single starting guess **x0** rather than an interval to search. That is, you can do `fzero(@myfunc, x0)`. In that case, **fzero** searches for an interval near **x0** where the sign of the function changes, and then looks for a root within that interval.

11.1.1 Something to watch out for

One of the biggest differences between R and MATLAB when using root-finding is the default tolerance. Let ϵ be machine epsilon; this is the smallest value you can add to 1.0 and get a value which is larger than 1.0 using the computer's representation of values. The value of ϵ is approximately 2.22×10^{-16} with double-precision floating point.[1] R uses a forward tolerance (i.e., change in x from one step to the next) of $\epsilon^{0.25} \approx 1.22 \times 10^{-4}$, while MATLAB uses a tolerance of ϵ. Depending on your function, this can produce striking differences in the estimates produced by the two packages. To tell R to use a tolerance of ϵ, you can use `uniroot(myfunc, c(a,b), tol=.Machine$double.eps)`. To change the tolerance in MATLAB to match the default behavior of R, use **optimset** to construct an optimization options set: `fzero(myfunc, [a b], optimset('TolX', eps^0.25))`.

11.2 Univariate optimization

To minimize a function $f(x)$ of one variable, as with root-finding, you should first define an R or MATLAB function which takes a single argument **x** and returns the function value $f(x)$ for that value of x. Suppose **myfunc** is such a function, and you wish to minimize it within the interval $a \leq x \leq b$. In some cases, you may wish your function f to receive extra parameters, i.e., define it as $f(x, p_1, p_2)$. These extra parameters could for example be data vectors, where the function computes a goodness-of-fit metric given the value of x. Here we will simply use example values of 17 and 121 for p_1 and p_2, respectively.

R

In R, use `optimize(myfunc, c(a,b))`. This returns a list with elements named **minimum**, containing the estimate $\hat{x}$ of the value of x which minimizes f, and **objective**, containing the value of the function there.

If your function receives extra parameters, you can pass them in by name to **optimize**, and it will pass them along to your function. For example, if your function **myfunc(x, p1, p2)** takes the parameter **x** to optimize over, along with two extra parameters, you can use a command like `optimize(myfunc, c(a,b), p1=17, p2=121)`.

To maximize a function, include the **maximum=TRUE** argument to **optimize**.

MATLAB

In MATLAB, use `fminbnd(@myfunc, a, b)` to obtain the estimate $\hat{x}$. To obtain other information, call **fminbnd** in a way similar to that used with **fzero** in Section 11.1: `[x,fval,exitflag,output]=fminbnd(@myfunc, a, b)`. This returns the estimate **x**, the function value **fval** there, an exit flag (which will be 1 upon success), and an **output** structure containing information about the number of iterations and function calls used.

If your function receives extra parameters, you can first construct an anonymous function (see Section 8.2.1) which receives only the argument **x**, but which then in turn calls your function with the extra parameters. For example, if your function **myfunc(x, p1, p2)** takes the parameter **x** to optimize over, along with two extra parameters, you can use a command like `fminbnd(@(x) myfunc(x,17,121), a,b)`.

---

[1]This value can be obtained via the command `.Machine$double.eps` in R, and `eps` in MATLAB.

To maximize $f(x)$, you should instead minimize the function $g(x) = -f(x)$. This can be done via a small anonymous function: `fminbnd(@(x) -myfunc(x), a,b)`.

11.3 Multivariate optimization

Minimizing a function $f(x, y)$ of several variables is something of a black art. If you visualize the function as describing a surface, there are various methods for trying to find a local minimum, but they can all behave very poorly for certain types of surfaces. They may become trapped in a local optimum, or converge very slowly if the value of the function changes much more quickly in some directions than others. Rather than a search interval or region, the standard methods here simply take a starting guess, and adaptively search "near" that initial point. As usual, you should write an R or MATLAB function implementing the mathematical function you are trying to minimize. Suppose the function f has two arguments, x and y. You should implement a function **myfunc(v)**, or **myfunc(v,p1,p2)** if the function also takes a couple of extra parameters that will not be part of the optimization. The first parameter to the function should be a vector **v**, whose elements are x and y, respectively. (The order of the actual function parameters within the vector **v** is not important, as long as you are consistent among your function implementation and your calls to the optimizer.) Also assume that the initial guesses will be $x = 1$ and $y = 2.2$, and again the extra parameters will have values $p_1 = 17$ and $p_2 = 121$ when they are provided.

R

Use `optim(c(1, 2.2), myfunc)`, which returns a list containing the following elements:

- **par**, a vector containing the estimates of the parameters in the same order in which they are packed into the vector parameter **v** of your function

- **value**, the value of the function at the value **par**

- **counts**, how many times your function was called

- **convergence**, which is either 0 (indicating success) or a code indicating a problem

- **message**, which is either **NULL** or a character string with some additional information.

If you just want the parameter values which minimize the function, you can extract just that element of the return value: `optim(c(1, 2.2), myfunc)$par`. To include extra parameters, provide them by name to **optim**, and it will pass them along to your function: `optim(c(1, 2.2), myfunc, p1=17, p2=121)`.

By default, the Nelder-Mead method is used, but you can request other methods; you can also provide a function which computes the gradient of your function if you have that information available.

To maximize or to change the stopping criteria, **optim** takes an additional parameter named **control**, which should be a list. If the list contains an element named **fnscale** which is negative, then the function will be maximized rather than minimized. You can also specify a value for **reltol** to provide a tolerance during the search (see the documentation for full details). For example, use `optim(c(1, 2.2), myfunc, control=list(fnscale=-1, reltol=1e-12))` to maximize a function with relative tolerance of 10^{-12}.

MATLAB

Use `fminsearch(@myfunc, [1 2.2])`, which returns a vector containing the estimates of the parameters in the same order in which they are packed into the vector parameter of your function. As with **fzero** and **fminbnd**, you can instead use `[x,fval,exitflag,output] = fminsearch(@myfunc, [1 2.2])` to retrieve additional information about the function value and whether or not the procedure terminated successfully. To include extra parameters for your function, you can again use an anonymous function as with the earlier numerical routines: `fminsearch(@(x) myfunc(x,17,121), [1 2.2])`.

11.4   Numerical integration

Both platforms include routines to perform numerical integration, or quadrature, using adaptive step sizes to achieve a desired level of accuracy. MATLAB also includes a function **trapz** which performs trapezoidal numerical integration. With a vector **y**, `trapz(y)` uses trapezoidal integration, assuming a corresponding set of x values with uniform unit spacing. You can use `trapz(x,y)` to provide your own set of corresponding x values. Note that **trapz** will also operate with matrices, performing the integration down each column, or along a specified dimension.

R does not include a built-in function to perform trapezoidal integration, but you can use a simple command to reproduce the effects of MATLAB's **trapz**. Given vectors **x** and **y**, use `sum(diff(x)*(y[-length(y)]+y[-1]))/2`. If you wish to assume the x values have unit spacing, you can omit the "**diff(x)***".

Now suppose you wish to integrate the function f over the interval from a to b using adaptive quadrature. In both R and MATLAB, your function **f** must be written in such a way that it can operate on an entire vector **x** of values simultaneously, and return the corresponding vector of function values.

In R, `integrate(f, a, b)` will work, using relative and absolute error tolerances of $\epsilon^{0.25}$. You can specify one or both tolerances via additional arguments: `integrate(f, a, b, rel.tol=tol1, abs.tol=tol2)`.

In MATLAB, `integrate(@f, a, b)` will work, using a default relative tolerance of 10^{-6} and absolute tolerance of 10^{-10}. You can specify different tolerances via `integrate(@f, a, b, 'RelTol', tol1, 'AbsTol', tol2)`. Note that the **quad** command was commonly used for numerical quadrature in previous versions of MATLAB, but it will be removed in a future release.

11.5   Curve fitting

11.5.1   Piecewise linear interpolation

Routines exist in both platforms to aid you in constructing a function which implements piecewise linear interpolation based on a set of points (x_i, y_i). Suppose **x** and **y** are vectors of the same length, and **xnew** is a vector of x coordinates where you would like to know the linearly interpolated y coordinates. In R, you can use `approx(x, y, xnew)`. In MATLAB, you can use `interp1(x, y, xnew)` or `interp1(x, y, xnew, 'linear')`. As the latter form suggests, MATLAB's **interp1** command is capable of doing other types of interpolation as well, including piecewise constant functions and cubic splines.

11.5.2 Polynomial fitting

Suppose you have vectors **x** and **y** containing the coordinates (x_i, y_i) of some points, and wish to fit a polynomial $y = p(x) = c_0 + c_1 x + c_2 x^2 + \cdots + c_n x^n$ to the data. The coefficients $c_0, c_1, \ldots, c_n$ can be chosen in the least-squares sense, to minimize the sum of squared residuals $\sum_i [y_i - p(x_i)]^2$.

R

The **lm** ("Linear Models") function can be used for this purpose, but it it is only convenient when the degree n of the polynomial is small (which it typically is). Note that each of the commands below returns a vector with the coefficients of the polynomial in ascending order, i.e., $c_0, c_1, \ldots$.

To fit a line $y = c_0 + c_1 x$, use `coef(lm(y ~ x))`. To fit a quadratic $y = c_0 + c_1 x + c_2 x^2$, use `coef(lm(y ~ x + I(x^2)))`.

To fit a polynomial of higher degree n, you can use a command of the above form, although it becomes tedious for large n. To automate the process, you can first construct a string of the form "**y ~ I(x ^ 1) + I(x ^ 2) + $\cdots$ + I(x ^ n)**", and then feed that string as a formula to the **lm** function. This can be done via the following somewhat daunting commands, which first build up a string containing the right-hand side of the formula (i.e., everything after the tilde):

```R
RHS = paste('I(x^', 1:n, ')', sep='', collapse='+')   # right-hand side
coef(lm(as.formula(paste('y~', RHS))))
```

If you wish to fit a polynomial with a zero intercept, you can include a "-1" in the formula. For example, to fit the function $y = c_1 x + c_2 x^2$, use `coef(lm(y ~ -1 + x + I(x^2)))`.

MATLAB

The **polyfit** function will conveniently fit a polynomial. Simply use `polyfit(x, y, n)` to fit a polynomial of degree **n**. The return vector has the coefficients in descending order, e.g., c_2, c_1, and c_0 for a quadratic fit obtained via `polyfit(x, y, 2)`.

To fit a polynomial with zero intercept, there is no simple way to use **polyfit** for the task. Instead, you can write a function which computes the sum of squared residuals given the data points and coefficients, and then minimize that function. For example, to fit a quadratic polynomial with zero intercept, $y = c_1 x + c_2 x^2$, you can first create a file **L.m**:

```matlab
function retval = L(v, x, y)
% Sum of squared residuals for quadratic polynomial with zero intercept
% The polynomial is c1*x + c2*x^2  where v(1)=c1 and v(2)=c2
c1 = v(1);
c2 = v(2);

yhat = c1*x + c2*x.^2;  % fitted values
retval = sum((y - yhat).^2);
```

Then, to perform least-squares optimization with starting guesses of 1 for both coefficients, use `fminsearch(@(v) L(v,x,y), [1 1])`.

11.5.3   Splines

Both R and MATLAB have a **spline** function capable of fitting cubic splines; consult the documentation, as both platforms' versions can compute a variety of versions of splines, depending on the data vectors and arguments passed in. Assume **x** and **y** contain some data points, and **sx** contains some x values where we want to know the values of the interpolated spline. In R, `tmp = spline(x, y, xout=sx)` uses the method of Forsythe, Malcolm, and Moler, where an exact cubic polynomial is fitted through the four points at each end of the data. It returns a list **tmp** containing **x** and **y** components; the **x** component simply contains the **sx** values, and the **y** component has the corresponding fitted y values. In MATLAB, `sy = spline(x, y, sx)` uses "not-a-knot" conditions, where the first two piecewise cubic polynomials coincide, as do the last two.

R's **spline** function can take a number of options to fit various types of splines. MATLAB has a Spline Toolbox which contains a number of functions that can fit splines; the **csapi** function is a good place to start.

11.6   Differential equations

Both platforms provide facilities for numerically solving systems of differential equations, using adaptive step sizes to achieve a desired level of estimated error. We will consider three cases as examples.

I. The single ordinary differential equation $dx/dt = 5x$, which we will solve from $t = 3$ to $t = 12$ with initial condition $x(3) = 7$.

II. The system of two predator-prey ODEs $dw/dt = 1.8w - 0.7wz$ and $dz/dt = 1.26wz - 0.6z$, from $t = 3$ to $t = 12$ with initial conditions $w(3) = 1$, $z(3) = 1.2$.

III. The single ODE $dx/dt = rx(1 - x/K)$ from $t = 0$ to $t = 20$ with initial condition $x(0) = 2.5$. Note that this ODE includes parameters r and K; we will use values $r = 1.3$ and $K = 50$.

To numerically solve ODEs in both platforms, you must write a function which receives the current time, a vector containing the current values of the state variables, and the values of any additional parameters in the ODEs. The function should return the rates of change of the state variables.

R

The function **lsoda** in the package **deSolve** is a commonly used method for numerically solving ODEs (see Section 13.9 for information about installing and loading packages). It automatically switches between nonstiff and stiff solvers.

The function you write receives three parameters:

t: the current time

y: a vector containing the values of the state variables at the current time

parms: a vector containing the values of any additional parameters.

Your function should return a list, whose first element is a vector containing the derivatives of the state variables. If your list contains additional elements, they are recorded as the values of "global" values over time. If you are coming from MATLAB, it can be easy to forget the requirement that your function must pack the vector of derivatives within a list.

To solve the ODE in the first example, use the function below.

```
———————————————— func1.R ————————————————
func1 = function(t, y, parms) {
  return(list(5*y))
}
```

Again, speaking from (too much) experience, a common mistake is to forget to put the return value inside of a **list**, but fortunately **lsoda** gives a clear error message if you do so.

To solve the ODE using equally spaced values between $t = 3$ and $t = 12$ with a spacing of $\Delta t = 0.1$, use `tmp = lsoda(7, seq(3,12,0.1), func1, NA)`. The parameters to **lsoda** are:

- The initial condition $x(3)$

- The sequence of time values for which you want the value of x

- The function defining your ODE

- Additional parameters to pass to your function (**NA** in this case, because no additional parameters are needed).

The **lsoda** function returns a matrix with class **deSolve**. The first column contains the time values you requested. The next set of columns contains the state variables at those times (in this example we only have one state variable). If your function returns any global values, those will be contained in subsequent columns.

Because the return value from **lsoda** is a special type of matrix, you can simply use `plot(tmp)` to plot x versus time. Or to be more explicit you can use `plot(tmp[,1], tmp[,2], type='l')`.

For the second example ODE, use the following function.

```
———————————————— func2.R ————————————————
func2 = function(t, y, parms) {
  w = y[1];   z = y[2]   # assign convenient names
  dwdt = 1.8*w - 0.7*w*z
  dzdt = 1.26*w*z - 0.6*z
  return(list(c(dwdt, dzdt)))
}
```

A common mistake to make here is to forget the **c()**, and return a list containing the two derivatives, rather than a list containing a single vector which has the two derivatives. If you do this, **lsoda** complains about the number of derivatives returned being incorrect.

The command `tmp=lsoda(c(1,1.2), seq(3,12,0.1), func2, NA)` numerically solves the ODEs. You can use `plot(tmp)` to make two separate subplots showing w versus time and z versus time, respectively. Or use `matplot(tmp[,1], tmp[,2:3], type='l')` to plot them together.

For the third example ODE, the method for writing your function and calling **lsoda** are very similar to the first example. You will pass a vector of parameters as the fourth argument to **lsoda**, rather than simply **NA**.

```
─────────────────────── func3.R ───────────────────────
func3 = function(t, y, parms) {
  r = parms[1];  K = parms[2]  # assign convenient names
  return(list(r*y*(1-y/K)))
}
```

Then solve via `tmp=lsoda(2.5, seq(0,20,0.1), func3, c(1.3,50))`.

Consult the documentation for **lsoda** for the wide variety of arguments it takes regarding tolerances, Jacobians, and so on.

MATLAB

There are a number of functions in MATLAB used for solving systems of differential equations. The different functions use solvers of different order, are specialized for stiff or nonstiff systems, and so on. Probably the most commonly used solver is **ode45**, a fourth- to fifth-order Runge Kutta solver for nonstiff systems.

The function you write requires two parameters, but can take more:

t: the current time

y: a vector containing the values of the state variables at the current time

additional parameters: optional parameters for the ODEs can be provided as well.

Your function should return a column vector containing the derivatives of the state variables. Note that returning a row vector will generate an error.

To solve the ODE in the first example, use the function below.

```
─────────────────────── func1.m ───────────────────────
function retval = func1(t,y)
  retval = 5*y;
```

To solve the ODE and let **ode45** choose the spacing between time intervals, use `ode45(@func1, [3 12], 7)`. The parameters to **ode45** are:

- A handle (see Section 8.2.7) to your function

- The beginning and end of the time interval to solve over

- The initial condition.

The **ode45** function by default will just plot the solution it calculates. If you wish to save the results it computes, use `[t,y] = ode45(@func1, [3 12], 7)`. The **t** vector will contain the time values, and the **y** vector the corresponding values of the state variable.

Rather than letting **ode45** decide the spacing of the time intervals, you can specify it yourself by providing a vector of time values instead of simply the beginning and ending points: `[t,y] = ode45(@func1, 3:0.1:12, 7)`.

For the second example ODE, use the following function.

```
─────────────────────── func2.m ───────────────────────
function retval = func2(t,y)
  w = y(1);  z = y(2);  % assign convenient names
  dwdt = 1.8*w - 0.7*w*z;
  dzdt = 1.26*w*z - 0.6*z;
  retval = [dwdt; dzdt];
```

The command `[t,y] = ode45(@func2, [3 12], [1; 1.2])` numerically solves the system. Leaving off the `[t,y] =` will cause **ode45** to automatically plot the results, or you can use the command as given followed by `plot(t, y)`. Also note that the vector of initial conditions can be provided as a row or column vector. Because the function must return the derivatives as a column vector, it may be helpful to be consistent and provide the initial conditions in column form as well.

For the third example ODE, use the following function.

```
                                     func3.m
function retval = func3(t,y,r,K)
  retval = r*y*(1-y/K);
```

Then solve via `[t,y]=ode45(@func3, [0 20], 2.5, [ ], 1.3, 50);`. Note that the empty array `[ ]` is there as a placeholder for the fourth parameter to **ode45**, which can contain various options controlling the behavior of the solver.

You can use the **odeset** function to create a set of options for **ode45**, such as tolerances, Jacobians, and so on. For example, to specify a relative tolerance of 10^{-4}, do the following: `[t,y]=ode45(@func3, [0 20], 2.5, odeset('RelTol', 1e-4), 1.3, 50)`. Consult the documentation for **ode45** and **odeset** for more options.

There are a variety of other solvers you can try if **ode45** has trouble, for example for a stiff system (one where different state variables are changing over very different time scales). If **ode45** fails, **ode15s** is usually a good one to try next. Others include **ode23**, **ode23s**, **ode23t**, **ode23tb**, **ode113**, and **ode15s**. See the documentation for more details about the various methods. In particular, the reference page for **ode45** brought up via the command `doc ode45` includes a table suggesting when to try which solver.

12

File Input and Output

Getting data into and out of files is one of the fundamental necessities of a computing platform. There are many facilities for doing so in both R and MATLAB.

For reading text data from files, R has the functions **read.table**, **scan**, and **readLines**, while MATLAB has **load**, **fgetl**, **fscanf**, **textread**, **textscan**, and **importdata**. Often, more than one of these functions can be used to accomplish a given task. Which one is best is partly personal preference, and may also depend on the type of variable you want to store the data in. The following examples of some common tasks may help to suggest some possible approaches. Note that these commands will all read and write files in the current working directory (see Section 7.1), unless you specify a path to a file in another directory.

12.1 Opening files

Both R and MATLAB have various functions which will operate on files via what are called *file descriptors*.[1] A file descriptor is a "handle" which refers to an open file, and generally appears to the user as an integer. In MATLAB, as in the C programming language, the file descriptors 0, 1, and 2 are generally reserved for standard input, standard output, and standard error, the standard input/output channels for programs. In R, these three connections are accessed via the function calls `stdin()`, `stdout()`, and `stderr()`.

When you open a file and create a file descriptor, you can specify whether you want to open the file for reading, writing, appending, or a combination of reading with writing/appending. This is done by providing a permissions mode string. The string can be "r," "w," or "a" for reading, writing, and appending, respectively. Using "r+," "w+," and "a+" specifies reading and writing without overwrite, reading and writing with overwrite (discarding the existing contents of the file), and reading and appending.

In R, you can open a file for reading via the command `fid = file('filename', 'r')`. You can test for success by calling `isOpen(fid)`, which returns **TRUE** if the file was successfully opened. In MATLAB, you can use `fid = fopen('filename', 'r')`. It will return a positive integer upon success, and -1 on failure. In MATLAB, the `'r'` string is optional, as that is the default.[2] Also in both platforms, be aware that if you open a file for writing (using either `'w'` or `'w+'`), the file will be overwritten, i.e., any existing data in them will be erased without warning! When you are done accessing a file, you should

---

[1] R uses what are called *connections* rather than simple file descriptors, but for most practical uses, they are equivalent.

[2] Technically the `'r'` string is optional in R as well. But note that if you omit it, actually opening the file will be deferred until you try to access the file (i.e., read from it). It will be closed again immediately after reading, and the current position for reading will be reset. That is, `fid=file('foo.txt'); isOpen(fid); x1=scan(fid,n=1); x2=scan(fid,n=1)` will show that **fid** is not open, and will then read the first value from the file twice, storing it in **x1** and **x2**. Including the `'r'` in the call to **file** will cause it to read the first two values on the consecutive calls to **scan**.

close its file descriptor. In R, you can do this via `close(fid)`, and in MATLAB, you can use `fclose(fid)`.

Many of the R functions for working with files can accept either file descriptors or file names; the examples given in this chapter are all written to use file names, rather than explicitly opening and closing the file descriptor (or the "connections" that encapsulate the file descriptors).

12.2   Reading a table of numbers

One of the simplest common tasks is to read a table of numbers from a text file. For example, say the file **mynumbers.txt** contains the following:

```
———————————————— mynumbers.txt ————————————————
4 8 15 16 23 42
8 10.4 7.7 14.2 5.9 6.1
12.3 17 8.5 11.1 21 18.3
```

It is also common to be given a file where the first line is a header containing the names of the columns, as below:

```
———————————————— numwithheaders.txt ————————————————
height weight length stress age thickness
4 8 15 16 23 42
8 10.4 7.7 14.2 5.9 6.1
12.3 17 8.5 11.1 21 18.3
```

You may even have a file which has both column labels at the top and row labels to the left:

```
———————————————— rowcollabels.txt ————————————————
height weight length stress age thickness
Troy 4 8 15 16 23 42
Ithaca 8 10.4 7.7 14.2 5.9 6.1
Cambridge 12.3 17 8.5 11.1 21 18.3
```

For the last file above, note that the first row has six fields, while subsequent rows have seven fields each.

R

To read the first file into an R matrix, you can use the command `A=as.matrix(read.table('mynumbers.txt'))`. Note that the matrix **A** will have column names **V1**, **V2**, **V3**, etc. You can remove those names via `colnames(A)=NULL` if desired. R will ignore comments prefixed by a hash mark (#). That is, the following text file will give the same results when read in using the above method:

```
———————————————— mynumbers-Rcomments.txt ————————————————
# This file contains the
# numbers I will be using
4 8 15 16 23 42  # lucky numbers
8 10.4 7.7 14.2 5.9 6.1     # some unlucky numbers?
12.3 17 8.5 11.1 21 18.3
```

By default, **read.table** requires the values in the file to be separated by white space (tabs or spaces). You can use the **sep** parameter to specify other separators between values. For example, if your values are separated by commas, you can use A=as.matrix(read.table('foo.txt', sep=',')). Also see **read.csv** and **read.csv2** for reading comma-separated values, or for semicolon-separated values which use commas as decimal points (but also take care that those functions default to **header=TRUE**, which treats the first line of the file specially, interpreting it as the names of the variables).

To read the files with column labels, use the command A = as.matrix(read.table('numwithheaders.txt', header=TRUE)) to inform **read.table** that the first row has header information. For the file with row and column labels, use A = as.matrix(read.table('rowcollabels.txt')). Because the first row contains one fewer field than the other rows, **read.table** deduces that the first row contains column labels and the first column contains row labels (but see the **row.names** parameter for how to specify otherwise).

Note that **read.table** is really designed for reading data frames, not matrices, and it can be slow for large files. The use of **scan** is encouraged for reading large matrices from files. However, **scan** will not automatically determine the size of the matrix; it will return all the values in a vector, which you can then convert into a matrix. For the simple data file above, you could use the following command to read it in: A = matrix(scan('mynumbers.txt'), ncol=6, byrow=TRUE). The **scan** function also allows you to use the **sep** parameter to specify the separator between values in the file. It will not skip comments by default, but you can use the **comment.char='#'** argument to skip over R comments in the file. If you are calling **scan** from within a function or script, you may also want to use the **quiet=TRUE** argument so that **scan** does not report how many items it read in.

For the file with a header line at the top, you can tell **scan** to skip over that first line via its **skip** argument: A = matrix(scan('numwithheaders.txt', skip=1), ncol=6, byrow=TRUE). See below for how to handle the third file with row and column labels.

As seen above, you can use **matrix** to reshape the vector returned by **scan** into a matrix. There are other ways to use **scan** to read in data and turn them into a matrix. First, you can use the **what** parameter to tell **scan** what types of values (and how many) are on each line. For example, to tell it there will be 6 numeric values on each line, you can let **what** be a list containing six numeric values, as in tmp = scan('mynumbers.txt', what=as.list(numeric(6))). This reads the data into a list of vectors; the ith element of the list contains the ith column of data as a vector. You then need to convert this result to a matrix; this can be done using the ideas from Section 5.5.2. Here, we can simply use either A = t(do.call(rbind,tmp)) or A = simplify2array(tmp). You can also use the **what** parameter with **scan** to name the columns of your data, by providing a list with the names you would like to use. The data types of the list elements specify the types of values being read in; you can simply use 0 to specify numeric data. So for example, you could use tmp = scan('mynumbers.txt', what=list(height=0, weight=0, length=0, stress=0, age=0, thickness=0)) to read the data into the list **tmp**, and then do the subsequent conversion to a matrix as above, to read in the data using the labels *height*, *weight*, *length*, *stress*, *age*, and *thickness* as the labels for the six columns of data.

The **what** parameter with **scan** also allows you to handle the third data file above, with row and column labels. You can build a **what** list for **scan** containing one string and 6 numeric values, and then convert the numeric data (in elements 2–7 of the list) to a matrix:

```
────────────────────────── R ──────────────────────────
mywhat = c(list(''), numeric(6))
tmp = scan('rowcollabels.txt', skip=1, what=mywhat)
A = simplify2array(tmp[2:7])
```

MATLAB

To read a simple text file with numeric data into a MATLAB matrix, you can simply do
`A=load('mynumbers.txt')`. The numbers within a row in the file can be separated by
spaces, tabs, commas, or semicolons. MATLAB will also ignore comments prefixed by a
percent sign (%). That is, the following text file will give the same results when read in
using **load**:

```
―――――――――――――― mynumbers-matlabcomments.txt ――――――――――――――
% This file contains the
% numbers I will be using
4 8 15 16 23 42  % lucky numbers
8 10.4 7.7 14.2 5.9 6.1    % some unlucky numbers?
12.3 17 8.5 11.1 21 18.3
```

The command `A=importdata('mynumbers.txt')` can also be used. This method, however,
will not work if there are MATLAB comments in the file, and spaces are the only delimiters
allowed (somewhat strange results are given if any of these conditions are violated).

There is no very simple way to skip over the header line in the other two files and read
the values into a matrix, but there are several ways to accomplish the task. You can first
open the file, manually read and discard one line using **fgetl**, and then read the data using
fscanf. By default, **fscanf** reads all data from the file into a single vector. You can provide
an argument indicating either the total number of elements to read in (the data will still
be read into a single vector), or the size of a matrix to place the data in. In the latter case,
the data across the rows of the files are placed down the columns of the matrix. You must
specify the number of rows in the matrix (i.e., the number of columns of data in the file).
If you do not wish to specify how many lines of data there are, specify an infinite number
of columns for the matrix. The return value of **fscan** can then be transposed so that the
matrix entries have the same orientation as the data in the file. The following commands
achieve this:

```
―――――――――――――――――――――――― MATLAB ――――――――――――――――――――――――
fid = fopen('numwithheaders.txt');
fgetl(fid);   % don't store return value, to discard first line of file
A = fscanf(fid, '%f', [6 inf])';  % for 6 columns of data, all rows
fclose(fid);
```

You could also use **textread**. It takes a filename to read from (rather than a file descriptor), a format string, and some other optional arguments. You can use one of the optional
arguments to indicate how many header lines to skip over at the top of the file. The command `v = textread('numwithheaders.txt', '%f', 'headerlines', 1)`, for example,
skips over one header line. This will read all 18 values into a single vector, which you can
then reshape into a matrix. Recall that reshaping a vector into a matrix places the elements down the columns of the matrix, so we should reshape into a 6×3 matrix and then
transpose: `A = reshape(v, 6, 3)'`.

For the third file, with row and column labels, yet another approach is best. If you try
to use **fscanf** with the **rowcollabels.txt** file, the combination of character and numeric
data on each row causes the entire matrix returned by **fscanf** to be converted to numeric
values. This is inconvenient, especially because the text labels on the rows have different
lengths, and so will take up varying numbers of elements in any vector or matrix returned
by **fscanf**. The function **textscan** works better here, as it returns its results in a cell array,
with one element per column of data from the file. You provide **textscan** with an **fprintf**-
style format string as well. All but the first element of the cell array, i.e., all of the columns
of numeric data, can then be converted into a standard matrix. Like **textread**, **textscan**

has many optional arguments, including one indicating how many header lines to skip over at the top of the file.

```
──────────────────────────── MATLAB ────────────────────────────
fid = fopen('rowcollabels.txt');
tmp = textscanf(fid, '%s%f%f%f%f%f%f', 'Headerline', 1);
fclose(fid);
A = cell2mat(tmp(2:end));
```

textread can also be used to read in the file with row and column labels, but you must provide as many output values as there are fields in the format string. Since there are seven columns (one string and six numeric values), seven output values should be provided: `[col1,col2,col3,col4,col5,col6,col7] = textread('rowcollabels.txt', '%s%f%f%f%f%f%f', 'headerlines', 1)`. The variable **col1** will then be a cell array containing the strings from column 1 of the file, while each of the other variables **col2** through **col7** will be a column vector containing data from the corresponding column of the file.

12.2.1 Subsets of a data file

Sometimes, when writing a program to analyze a large data file, it is convenient to work with a smaller subset of the data while debugging your script. Once the script is working, you can then apply it to the entire data set. Although you could in theory simply read in the entire data file and then extract just a portion of the data matrix to work with for debugging, it can be faster to read in only part of your data file, especially if the file is very large.

R

The **read.table** function takes an argument **nrows** to limit the number of rows it reads. For example, you can do `A = as.matrix(read.table('mynumbers.txt', nrows=2))` to read in just the first two lines of the data file.

Similarly, **scan** takes an argument **n** indicating the maximum number of elements to read in, as in `A = matrix(scan('mynumbers.txt', n=12), ncol=6, byrow=TRUE)`. When **scan** is called in this way, the argument **nmax=12** would have the same effect.

When you provide **scan** with a list for its **what** parameter, **nmax** instead indicates the maximum number of records to read, i.e., how many times to apply the **what** list.[3] So this will also limit the data read in to the first two lines of the file: `tmp = scan('mynumbers.txt', what=as.list(numeric(6)), nmax=2); A = simplify2array(tmp)`.

MATLAB

Neither **load** nor **importdata** provide a way to restrict the number of rows read in. But the **fscanf** function, which lets you provide an argument indicating how many elements to read in, will work. As stated before, **fscanf** reads the data from the rows of a file into the columns of a matrix, so if you wish to read in 2 rows with 6 columns each, you should specify that you want a 6×2 matrix and then transpose the resulting matrix to orient it the same way as the data file:

---

[3]The **what** list does not necessarily correspond to one entire line. The **what** list could get used multiple times for a single line, or it may take multiple lines to gather enough data to match the entire **what** list.

```
─────────────────────── MATLAB ───────────────────────
fid = fopen('mynumbers.txt');
A = fscanf(fid, '%f', [6 2])'
fclose(fid);
```

If you instead used `fscanf(fid, '%f', 12)`, the 12 elements from the first two rows of the data file would be read into a 12×1 column vector.

12.3   Reading numeric data with a different comment character

Sometimes one has a file containing numbers with comments delimited by a different comment character. This happens, for example, if you want to read a file that was intended for MATLAB (and has its comments prefixed by "%" characters) into R, or vice versa. Usually I will simply open a text editor and use search-and-replace to fix the comment characters to match whichever software platform I am using, but sometimes (particularly if there are many files to read), this is not convenient.

R

This is very simple to do in R, thanks to the **comment.char** argument to the **read.table** function. You can simply do `A=as.matrix(read.table('mynumbers-matlabcomments.txt', comment.char='%'))`. The **scan** function also allows you to specify a **comment.char** parameter.

MATLAB

None of **load**, **importdata**, and **fscanf** have a way to specify a comment character, but **textscan** does, through its optional **CommentStyle** parameter. **textscan** will return the data in a single vector (which is contained in the first element of a cell array), which can be reshaped into a matrix:

```
─────────────────────── MATLAB ───────────────────────
fid = fopen('mynumbers-Rcomments.txt');
tmp = textscan(fid, '%f', 'CommentStyle', '#');
fclose(fid);
A = reshape(tmp{1}, 6, 3)'  % convert to matrix
```

If you want a way to accomplish the task without knowing in advance how many columns the file has, you can read the file in a line at a time using **fgetl**, and then use **textscan** to parse the data from each line. This can be done as follows:

```
─────────────────────── MATLAB ───────────────────────
fid = fopen('mynumbers-Rcomments.txt');  % open the file
A = [ ];  % initialize empty matrix
while 1
  tline = fgetl(fid);  % read next line
  if ~ischar(tline), break, end  % exit loop if end-of-file
  tmp = textscan(tline, '%f', 'CommentStyle', '#');
  v = tmp{1};  % extract cell element as a vector
  % The line may have only contained comments.
```

```
    % If we got any actual data, append to bottom of matrix A.
    if (length(v) > 0), A = [A ; v(:)']; end
  end
  fclose(fid);
```

Note that **textscan** can process text from a string variable (**tline** above) as well as via a file-descriptor. The above script dynamically grows the matrix **A**, which will be very slow if the file is very large. If you have a large file, you may want to preallocate the correct amount of space for **A** if you know how many lines containing data are in the file, and then fill them in rather than appending to **A**, or dynamically grow the size of the matrix in a more clever way than just one row at a time. Another possibility would be to scan the file to count how many lines of data there are but without storing anything, preallocate space for the matrix **A**, and then read the file again while storing the data.

12.4 Reading numbers from a file where different lines have varying numbers of values

Sometimes, different lines in a file may contain different numbers of values. For example, consider the following file:

──────────────── varyinglength.txt ────────────────
```
8 6 7
5 3 0 9
3 1 4 1
17 121
8
4 8 15 16 23 42
10 9 8
```

R

If you just wish to read all of the values into a single vector, **scan** is the best approach. By default, v=scan('varyinglength.txt') will read all of the values, line by line, into a single vector.

Another way to read and store the data would be to create a list, where each element of the list is a vector containing the values from a single line. This can be done by first using **readLines** to read everything into a list of strings (with each string containing a line of the file), and then converting to a list of numeric vectors using the method from Section 5.5.4:

──────────────── R ────────────────
```
tmp = readLines('varyinglength.txt')
mylist = lapply(tmp, function(x) scan(text=x, quiet=TRUE))
```

A third way to read the values in is to read them into a matrix, with each row of the matrix containing values from one line of the file, and padding the shorter rows with the special value **NA**. This can be done by using the **fill=TRUE** argument with **read.table**. However, the **varyinglength.txt** example file here highlights two dangers to watch out for when using **read.table**, which can be seen in the output of the command below:

──────────────── R ────────────────
```
> A=as.matrix(read.table('varyinglength.txt',fill=TRUE))
```

```
> A
    X8 X6 X7
5    3  0  9
3    1  4  1
17 121 NA NA
8   NA NA NA
4    8 15 16
23  42 NA NA
10   9  8 NA
```

First, **read.table** determines the upper bound on the number of entries per row by examining only the first five lines of the file. The first five lines in this example file have at most four entries per line, so **read.table** assumed that all lines do. Because line 6 has six entries, its values get spread out over two rows of the matrix (the last row containing the values from line 6 gets padded with **NA**). Second, because the first line of the file has three entries, which is one less than the maximum of four entries seen in the first five lines, **read.table** assumed that column 1 contains row labels and that line 1 contains labels for the remaining three data columns (observe the **X8**, **X6**, and **X7** labels at the top, and the fact that the first column of the displayed output has left-justified formatting; those are row labels rather than actual matrix entries). Our resulting matrix is 7×3, rather than 7×6 as it should be.

The first problem can be addressed by telling **read.table** the maximum number of data entries per row. This can be determined using the **count.fields** function, which counts the number of fields in each line of a file, and computing the maximum among those values. This number can then be used to build a set of column names, which is the only way to override **read.table**'s mechanism for automatically determining the number of columns (you cannot simply give a parameter saying there are 6 columns). The second problem can be fixed by explicitly specifying **header=FALSE** to let **read.table** know that the first line of the file contains actual data rather than column labels. So the following commands will read the file into a matrix, with shorter rows padded with **NA**:

```
―――――――――――――――――― R ――――――――――――――――――
maxCols = max(count.fields('varyinglength.txt'))
colNames = seq_len(maxCols)
A = as.matrix(read.table('varyinglength.txt', fill=TRUE, header=FALSE,
col.names=colNames))
```

That produces the desired matrix:

```
      X1  X2 X3 X4 X5 X6
[1,]   8   6  7 NA NA NA
[2,]   5   3  0  9 NA NA
[3,]   3   1  4  1 NA NA
[4,]  17 121 NA NA NA NA
[5,]   8  NA NA NA NA NA
[6,]   4   8 15 16 23 42
[7,]  10   9  8 NA NA NA
```

MATLAB

If you just wish to read all of the values into a single vector, **textscan** will do the trick:

```
────────────────── MATLAB ──────────────────
fid = fopen('varyinglength.txt');  % open file
tmp = textscan(fid, '%f');   % read floating-point values
fclose(fid);  % close file
v = tmp{1};  % extract data vector from cell array
```

The values could also be read into a cell array, where each element of the cell array contains a vector holding the values from one line of the file. This can be done by repeatedly calling **fgetl** to read individual lines from the file, using **textscan** on that string to extract the values, and then appending to a growing cell array. This can be done as follows:

```
────────────── matlabreadvaryinglines.m ──────────────
fid = fopen('varyinglength.txt');  % open the file
myData = { };  % initialize empty cell array
dataLines = 0;
while 1
  tline = fgetl(fid);  % read next line
  if ~ischar(tline), break, end  % quit if end-of-file
  tmp = textscan(tline, '%f');
  % The line may have only contained comments.
  % If we got any actual data, append to cell array
  if (length(tmp{1}) > 0)
    dataLines = dataLines + 1;
    myData{dataLines} = tmp{1};
  end
end
fclose(fid);
```

The above script dynamically grows the cell matrix, which is slow. If you know roughly how many rows of data your file has, you should preallocate space for them, or grow the cell array by more than one element at a time.

If you wish to read the values into a matrix, padding the rows corresponding to lines with fewer values, A = dlmread('varyinglength.txt') will work, but it pads the rows with zeros. These zeros are indistinguishable from actual zero values at the end of a line. To pad with another more distinctive value, such as **NaN**, you can first read the data into a cell array as above, where each cell contains one line's data. The cell array can then be converted to a padded array. The following commands will do this, assuming the **matlabreadvaryinglines.m** script above has already been run to set up **myData** and **dataLines**:

```
────────────── matlabreadvaryinglines2.m ──────────────
lineLengths = cellfun('length', myData);  % lengths of lines
maxLen = max(lineLengths);  % largest number of values on one line
A = NaN(dataLines, maxLen);  % create matrix full of NaN values
for i=1:dataLines
  A(i,1:lineLengths(i)) = myData{i};  % copy row of data into matrix
end
```

12.5   Reading numbers and strings

Now suppose we have some files which contain both numeric data and text strings that we wish to read in. We will consider a few cases, using the following example data files:

```
───────────────────────── numtext1.txt ─────────────────────────
Yellow   4.2   6.6
Green    9.2   0.4
Yellow   7.9   8.5
Blue     9.6   9.3
```

```
───────────────────────── numtext2.txt ─────────────────────────
Color BoneDensity Alertness
Yellow   4.2   6.6
Green    9.2   0.4
Yellow   7.9   8.5
Blue     9.6   9.3
```

```
───────────────────────── numtext3.txt ─────────────────────────
Female   9.6   Yellow   4.2   6.6
Male     4.9   Green    9.2   0.4
Male     8.0   Yellow   7.9   8.5
Female   1.4   Blue     9.6   9.3
```

```
───────────────────────── numtext4.txt ─────────────────────────
Gender Length Color BoneDensity Alertness
Female   9.6   Yellow   4.2   6.6
Male     4.9   Green    9.2   0.4
Male     8.0   Yellow   7.9   8.5
Female   1.4   Blue     9.6   9.3
```

R

read.table can handle all four of the example files given above, with a bit of assistance. `read.table('numtext1.txt')` and `read.table('numtext3.txt')` will work, reading the data into a data frame. Note that if you use **as.matrix** to convert the data frame to a matrix, the matrix will contain character data. All of the numerical values will be converted to their string representations. For the other two files, you must instruct **read.table** to skip over the first header line, e.g., `read.table('numtext4.txt', header=TRUE)`.

You can also use **scan** to read in the data files. You just need to provide a **what** list containing the appropriate data types for each line. Doing `tmp = scan('numtext2.txt', what = list(color='', bdensity=0, alertness=0), skip=1)` will return a list with elements named **color**, **bdensity**, and **alertness**. Each element is a vector containing that column's data:

```
───────────────────────────── R ─────────────────────────────
> tmp = scan('numtext2.txt', what = list(color='',bdensity=0,alertness=0),
  skip=1)
Read 4 records
> tmp
$color
[1] "Yellow" "Green"  "Yellow" "Blue"
```

```
$bdensity
[1] 4.2 9.2 7.9 9.6

$alertness
[1] 6.6 0.4 8.5 9.3
```

You can produce a matrix containing the numeric data from the file via the command
`A=simplify2array(tmp[2:3])`. The list **what** above contains an empty string and two
zero values, indicating that we want to read one string and two (double-precision) numeric
values. If you want to read a column of a file specifically as integer data, you can include
the value **as.integer(0)** in your **what** list. And of course the **nmax** argument could always
be used to limit how many lines of data are read in by **scan** (or more precisely, how many
times the **what** list is applied).

MATLAB

The **importdata** function can handle these data files, though some of them a bit oddly.
The command `A = importdata('numtext1.txt')` produces a structure with three fields,
as follows:

```
——————————————— MATLAB ———————————————
>> A = importdata('numtext1.txt')
A =
          data: [4x2 double]
      textdata: {4x1 cell}
    rowheaders: {4x1 cell}
>> A.data
ans =
    4.2000    6.6000
    9.2000    0.4000
    7.9000    8.5000
    9.6000    9.3000
>> A.textdata
ans =
    'Yellow'
    'Green'
    'Yellow'
    'Blue'
```

As shown above, **A.data** is an array containing the numerical data from the file, while
A.textdata is a cell array containing the string data. Although not shown above,
A.rowheaders contains the same strings as **A.textdata**, because **importdata** interprets
that first column of text data as row labels for the other numeric data columns.

The command `A = importdata('numtext2.txt')` produces a structure **A** with two
fields: **A.data** will be identical with the outcome above, while **A.textdata** will still contain
all of the character data, as shown below:

```
——————————————— MATLAB ———————————————
>> A.textdata
ans =
    'Color'      'BoneDensity'      'Alertness'
    'Yellow'     ' '                ' '
    'Green'      ' '                ' '
```

```
    'Yellow'      ' '                    ' '
    'Blue'        ' '                    ' '
```

Note that there will be no **A.rowheaders** — it is only created if there are row headers but no column headers. Similarly, **A.colheaders** is created if there are column headers but no row headers. **A.textdata** will be created if any kind of text data are read in.

Somewhat odd results are obtained from A = importdata('numtext3.txt'). **A.data** will still be a 4×2 matrix, even though this file contains a third column of numbers, and **A.textdata** contains the following:

```
─────────────────────── MATLAB ───────────────────────
>> A.textdata
ans =
    'Gender'     'Length'     'Color'      'BoneDensity'     'Alertness'
    'Female'     '9.6'        'Yellow'     ' '               ' '
    'Male'       '4.9'        'Green'      ' '               ' '
    'Male'       '8.0'        'Yellow'     ' '               ' '
    'Female'     '1.4'        'Blue'       ' '               ' '
```

This happens because **importdata** finds the rightmost column containing text data and treats that column, as well as all columns to the left of it, as text data. This happens even if some columns further to the left of that column contain only numeric data.

Assuming you know how many columns your data file contains and what types of data they contain, **textscan** is a better way to read in its contents. Our **numtext4.txt** file contains five columns, of type string, float, string, float, and float. We can read all of the data as follows:

```
─────────────────────── MATLAB ───────────────────────
>> fid = fopen('numtext4.txt');
>> tmp = textscan(fid, '%s%f%s%f%f', 'Headerlines', 1)
tmp =
  {4x1 cell}    [4x1 double]    {4x1 cell}    [4x1 double]    [4x1 double]
>> fclose(fid);
>> tmp{1}
ans =
    'Female'
    'Male'
    'Male'
    'Female'
>> tmp{2}
ans =
    9.6000
    4.9000
    8.0000
    1.4000
>> tmp{3}
ans =
    'Yellow'
    'Green'
    'Yellow'
    'Blue'
```

(For brevity, the fourth and fifth elements of **tmp** are not shown.) Element **i** of the cell array **tmp** contains the data from column **i** of the file. Also note that we instructed **textscan** to skip over the first line of the file, which contained headers/labels. If you wish to pack the numeric data (which are stored in elements 2, 4, and 5 of the cell array **tmp**) into a matrix, you can use the command `A = cell2mat(tmp([2 4 5]))`. If you only want to read in part of your data, say the first 2 lines, **textscan** takes an optional argument after the format string indicating how many times to use that format; `tmp = textscan(fid, '%s%f%s%f%f', 2, 'Headerlines', 1)` will do the trick.

12.6 Reading the raw character data in, a line at a time

R

The contents of a file can be read into a vector of strings, with each element of the vector containing one line of the file. The function **readLines** is designed just for this, for example: `s = readLines('numtext4.txt')`. To read in only the first three lines of the file, you can use `s = readLines('numtext4.txt', n=3)`.

MATLAB

The contents of a file can be read into a cell array of strings, with each string containing one line of the file. This can be done using **textscan**, instructing it to treat each line as a single string by specifying a newline as the delimiter:

```
——————————— MATLAB ———————————
fid = fopen('numtext4.txt');
tmp = textscan(fid, '%s', 'Delimiter', '\n');
fclose(fid);
tmp = tmp{1}  % the actual cell array of strings
```

Note that **textscan** returns a cell array, whose first element is itself a cell array containing the lines of the file. To read in only the first three lines of the file, you can change the second line of the code above to `tmp = textscan(fid, '%s', 3, 'Delimiter', '\n');`.

12.7 Writing a table of numbers

Now suppose you have a vector **v** or matrix **A**, and would like to write its contents to a file, either to read in again later or perhaps to share with another software application.

R

To write a vector out to a file, you can use `write.table(v, 'foo.txt', row.names=FALSE, col.names=FALSE)`. The same method works with a matrix: `write.table(A, 'foo.txt', row.names=FALSE, col.names=FALSE)`. If you use **write.table** with a vector, it outputs one value per row, i.e., it outputs it as a column vector. If you wish to output all of the values on one line, you can use **matrix(v, nrow=1)** to convert the vector to a matrix with one row.

Maximal precision is used when writing floating-point numbers to a file, but note that the limits of finite precision are different for the base-10 numbers written to a text file and the base-2 internal representation used by R. For example, observe the following:

```R
v = 1/3
write.table(v, 'foo.txt', row.names=FALSE, col.names=FALSE)
v2 =scan('foo.txt')
print(v2-v)
[1] -3.330669e-16
```

The difference of roughly 3×10^{-16} arises because of this discrepancy in representation.

You can control the format of saved values by first using **format** to convert the data to strings, and then writing those results. For example, to print values with five significant digits, you can do the following:

```R
> A = matrix(c(pi,5,exp(1),sqrt(2)),nrow=2,byrow=TRUE)
> tmp = format(A, digits=5, drop0trailing=TRUE)
> write.table(tmp, row.names=FALSE, col.names=FALSE, quote=FALSE)
3.1416 5
2.7183 1.4142
```

In the above commands, the **drop0trailing** argument to **format** instructs it not to pad all of the values with trailing zeros, i.e., to display **5** rather than **5.0000**. When a filename is not provided to **write.table** (which then uses the default argument **file = ''**), the output is displayed in the R console as shown above. The **quote=FALSE** argument is needed because **format** returns the formatted values in a string vector, and **write.table** by default puts quotes around all strings that it displays.

write.table has various options which may be useful. For example, the **sep** parameter specifies the separator value between entries on a single line. Use **sep=','** to separate the values with commas.

MATLAB

You can use the **save** function to output a vector or matrix. For example, `save('foo.txt', A, '-ascii')` will write the matrix **A** to the file **foo.txt**, with spaces separating the values on a given line. However, **save** does not have much flexibility. For example, there is no way to control the number of significant digits in outputted values or the separator between consecutive values on a single line of the output file.

Another method with more flexibility is **dlmwrite**. By default, commas are used to separate values on a line. To output **A** using spaces between values on a line, use `dlmwrite('foo.txt', A, 'delimiter', ' ')`. You can also control the format of the output; the default is to use 5 significant digits. To use 9 significant digits, include the **'precision', 9** arguments to **dlmwrite**. You can also use format strings from C, e.g., **'precision', '%6.3f'** specifies to use a minimum field width of 6, with 3 digits after the decimal point.

12.8   Writing a set of strings

You may have a set of strings you wish to output to a file; examples in this section provide ways to do so.

R

If you have a vector of strings, **sv**, you can use the command `write(sv, 'foo.txt')` to output the strings (one per line) to the file **foo.txt**.

If you instead have the strings stored in a list **sl**, you can output them by first temporarily converting **sl** to a vector: `write(sapply(sl, c), 'foo.txt')`.

MATLAB

The standard way to store a set of strings of different lengths in MATLAB is to keep them in a cell array. If you wish to display the strings in a cell array named **sc**, first create this helper function which outputs a string followed by a newline to the specified file-descriptor:

```
———————————— fprintfnl.m ————————————
function fprintfnl(s,fid)
fprintf(fid,'%s\n', s);
```

Then, use the following command to print to the file-descriptor **fid**: `cellfun(@fprintfnl, sc, num2cell(fid*ones(size(sc))))`.

The ugliness in the final argument is because **cellfun** does not provide a way for you to include a scalar parameter to pass along to **fprintfnl** every time it is called. So instead we must construct a cell array the same size as **sc**, each element of which contains the scalar value **fid**.

A simpler alternative is to simply iterate over the elements of the cell array, outputting one string at a time:

```
———————————— MATLAB ————————————
for i in 1:numel(sc)
  fprintf(fid, '%s\n', sc{i})
end
```

12.9   Saving and loading variables in binary format

Sometimes it is useful to save some or all of your variables in a format specific to the platform you are using (R or MATLAB), so that they can be reloaded later. This is useful if you need to shut down your computer while in the middle of something and then later resume where you left off, or perhaps to switch to another computer.

R

You can use `save(A, v, file='my2vars.RData')` to save the variables **A** and **v**. If you have the names of the variables you want to save stored in a vector of strings, you can use the **list** parameter instead, e.g., `save(list=c('A', 'v'), file='my2vars.RData')`. If you want to save all of your variables, you can use **ls** to obtain all of their names: `save(list=ls(all=TRUE), file='allmyvars.RData')`. Calling `save.image()` is a shortcut for saving all variables to the file **.RData**, as happens if you quit R and tell it you wish to save the workspace. You can specify another filename via a command like `save.image('myfile.RData')`.

You can load in variables saved as above via a command like `load('my2vars.RData')`.

load invisibly returns a vector containing the names of the variables loaded. You can see it by doing `print(load('my2vars.RData'))`.

MATLAB

You can use `save('my2vars', 'A', 'v')` to save the variables **A** and **v** to the file named **my2vars.mat** (MATLAB automatically appends the ".mat" suffix to the filename if it does not already have an extension). You can also use the command form, i.e., `save my2vars A v`. To save all variables to the file **allmyvars.mat**, use `save('allmyvars')`, or `save allmyvars`. Entering just the command **save** saves all variables to the file **matlab.mat**.

To load in variables saved as above, use `load('my2vars')`. If you assign the return value to a variable, it will be a struct containing the variables as the fields. For example:

```
──────────────────────── MATLAB ────────────────────────
>> s = load('/tmp/msave')
s =
    A: [3x6 double]
    v: [3.1416 5 2.7183 1.4142 0.9000 1.2346e+03]
```

After doing the above, you can obtain a list of the variables loaded by entering `fieldnames(s)`. Also, entering just the command **load** (with no arguments) attempts to load variables from the file **matlab.mat**.

12.10   Images

R

To read a JPEG image, you can use `img = readJPEG('foo.jpg')`. Note that **readJPEG** is part of the **jpeg** package. If the JPEG image has a resolution of $x \times y$ pixels, it will be read into a $y \times x \times 3$ array, where the three layers contain the red, green, and blue values (ranging between 0 and 1). The **rasterImage** function can be used to add the image to a specified region within an existing plot; you should first set up an empty plot if you do not already have an existing plot. For example, `plot(0:1, 0:1, type='n'); rasterImage(img, 0, 0, 1, 1)`. To save an image stored in a $y \times x \times 3$ array **A** in a JPEG file, you can use `writeJPEG(A, 'foo.jpg')` (also from the **jpeg** package).

For reading and writing GIF images, see the **read.gif** and **write.gif** functions in the **caTools** package.

MATLAB

To read a JPEG image, you can use `img = imread('foo.jpg')`. If the JPEG image has a resolution of $x \times y$ pixels, it will be read into a $y \times x \times 3$ array, where the three layers contain the red, green, and blue values (ranging between 0 and 255). You can use `image(img)` to display the image. To save an image stored in a $y \times x \times 3$ array **A** in a JPEG file, you can use `imwrite(A, 'foo.jpg', 'jpeg')`.

imread and **imwrite** can also handle GIF images, as well as a variety of other formats.

12.11 URLs

It may be desirable to import data directly from a Web site. Both platforms have facilities for doing this easily by reading from URLs (uniform resource locators), i.e., Web addresses.

R

You can first open a connection to a URL, rather than a file. For example, you can do `fid = url('http://www.example.com')`. After that, the various routines for reading data from files can be used with the opened connection. To read all of the data from the Web site into a vector of strings, you can use `s = readLines(fid)`. The connection can then be closed via `close(fid)`.

MATLAB

You can read everything from a given URL into a string variable via a command like `s = urlread('http://www.example.com')`. You can then extract data from that string by using **textscan**, which will accept a string as its first argument in place of a file descriptor. If you prefer to use **fscanf** rather than **textscan** to read data, you should instead use **sscanf**, which is the equivalent function that reads from strings rather than via file descriptors.

12.12 Excel files

R

Although there are ways to read Excel files directly from within R, the general advice is that it is perhaps best to avoid doing it [25]. It can be simpler to have Excel export the data in a text file, such as a **.csv** (comma-separated values) file. The OpenOffice program may be able to read an Excel file and allow you to export it as a csv file if you do not have access to Excel itself.

The **gdata** package includes a **read.xls** function which can read files from some versions of Excel, although it relies on Perl being installed on your system. The **xlsx** package includes functions for reading and writing Excel files, as well as for manipulating their contents. This includes being able to edit attributes such as colors, fonts, and data formats, and the ability to add or remove rows, sheets, etc. There are other options for users of the 32-bit version of R on Windows, and various options to write Excel files; consult Reference [25] for current information.

MATLAB

There is a function **xlsread** which allows you to read data from an Excel file. For full functionality, including the ability to read just ranges of cells from specified sheets in the file, you must be running under Windows and have Excel installed. Otherwise, some more basic functionality is provided. There is also a corresponding **xlswrite** function to output data in Excel format.

13

Miscellaneous

This chapter collects some other miscellaneous information, such as material about working with variables and strings.

13.1 Working with variables

Various methods of listing information about variables and clearing or deleting variables are provided below.

1. Display a concise list of defined variables.

R	MATLAB
`ls()`	`who`

2. Display a concise list of all defined variables whose names contain "xyz".

`ls(pattern='xyz')`	`who *xyz*`

(R): The argument to **ls** is a regular expression; see `help('regex')` for more.

(MATLAB): In addition to using the wildcard character "*," you can use regular expressions to search for variables via a command like `who -regexp m[^p]`; see `doc regexp` for more.

3. Display a more detailed list of all defined variables.

`ls.str()`	`whos`

The same mechanisms for pattern matching using regular expression from the previous item can be used here as well.

4. Display detailed information about the variable **foo**

`str(foo)`	`whos foo`

5. Display detailed information about all variables whose names contain the string "xyz"

`ls.str(pattern='xyz')`	`whos *xyz*`

6. Open a graphical data editor to edit the value of variable **A**. (This can be especially useful for editing values within a matrix.)

`fix(A)`	`openvar A`

7. Remove/clear/delete the variable **x**.

`rm(x)`	`clear x`

8. Remove/clear/delete the two variables **x** and **y**.

rm(x,y)	clear x y

9. Remove/clear/delete all variables.

rm(list=ls())	clear all

10. Query to see if the variable **x** exists.

exists('x')	exist('x')

Both of these functions can take additional parameters which allow you to search for particular types of things, such as functions, or MATLAB M-files in the search path, etc.

13.2   Character strings

Suppose you have a string variable, created in either platform by a command such as s = 'Hello, world!', and you wish to display the contents of the string.

In R, you can use the command print(s). However, the output is a bit ugly:

```
──────────────────────── R ────────────────────────
> print(s)
[1]  "Hello, world!"
```

You can hide the quotes by using print(s, quote=FALSE) (or print(noquote(s)) will also work), but the "[1]" will still be there in the output. To avoid that, you can instead use cat(s). There is still one more problem, however. If you enter the commands cat(s); cat(s), you will notice that they get concatenated on the same line. You can fix that by instead using cat(s, fill=TRUE). Or instead, you can include C-like character escapes in your string; for example, cat('Hello!\n'). The **cat** function can actually take a vector of strings and display all of them; by default, they are separated by a single space, but you can specify a different separating string instead: v=c('Hi', 'there!'); cat(v, sep='++'). Finally, you can also use the **sprintf** command to build a string using C-style formatting codes, as shown below.

```
──────────────────────── R ────────────────────────
name = 'Alice'
numGoals = 12
cat(sprintf('%s scored %d goals', name, numGoals), fill=TRUE)
```

Things are much simpler in MATLAB. To display the string in variable s, you can simply use disp(s). The **disp** function will append a newline, so if you call it twice, the two strings will appear on separate lines. If you do not want a trailing newline, for example if you want to print two strings on the same line, use fprintf(s) instead. You can also use **sprintf** to build a string using C-style formatting codes, as shown below.

```
─────────────────────── MATLAB ───────────────────────
name = 'Alice';
numGoals = 12;
disp(sprintf('%s scored %d goals', name, numGoals))
```

For the **sprintf** routine in both platforms, use %d to refer to decimal integers, %f for floating-point numbers, **%e** for scientific-notation floating point, and **%g** to automatically choose **%f** or **%e** based on the value. You can specify field widths or precisions, such as **%5d** for decimal integers padded to 5 spaces, or **%.7f** for floating-point with 7 digits of precision. Consult the documentation for more information about the many formatting options accepted by **sprintf**.

Various other manipulations involving strings are shown below.

11. Create a string containing an apostrophe, such as "It's nice".

R	MATLAB
`s='It\'s nice' or s="It's nice"`	`s = 'It''s nice'`

12. Concatenate two strings **s1** and **s2**.

`paste(s1, s2)`	`[s1 s2]`

13. Concatenate a set of strings stored in **v**.

`v=c('first ', 'second');` `paste(v, collapse='')`	`v={'first', 'second'};` `strcat([v{:}])`

(MATLAB): If you just write `strcat(v{:})`, trailing spaces on strings will be dropped.

14. Extract characters 2–6 from the string in **s**.

`substr(s, 2, 6)`	`s(2:6)`

15. Find the indices of locations where the regular expression **p** appears within string **s**.

`inds = gregexpr(p,s)[[1]]`	`inds = regexp(s,p)`

(R): If you call `gregexpr(p,v)` where **v** is a vector of strings, then the return value will be a list, where element **i** of the list contains information about the matches of the pattern **p** against element **i** of the vector **v**. When called with a single string **s**, the return value is a list with one element. Each element of the returned list has additional attributes set (indicating the lengths of the matches found, and information about the character encoding). You can remove those attributes via `attributes(inds)=NULL`.

(MATLAB): If one of **s** and **p** is a string and the other is a cell array of strings, then the matching will be performed using the string together with each element of the cell array. A cell array will then be returned. If both **s** and **p** are cell arrays, then they should be the same length, and pairwise matches will be performed (with the results returned in a cell array).

16. If you wish to test whether a regular expression matches against a string, in MATLAB, you can still use **regexp** as above, because an empty vector evaluates as FALSE, and a vector containing any indices of matches evaluates as TRUE. However, in R, you should use a different function.

`if (grepl(p,s)) {`
`    ...commands...`
`}`
	`if (regexp(s,p))`
	`    ...commands...`
	`end`

17. Convert a number to a string.

`as.character(x)`	`num2str(x)`

Note that you can also use **sprintf**. For example, in both platforms, the command sprintf('%.5f', pi) will work.

18. Compare two strings to determine whether they are the same.

s1 == s2	strcmp(s1,s2)

(MATLAB): The **strcmp** function can also handle either one or both of **s1** and **s2** being cell arrays. In that case, the cell arrays are handled in a manner similar to that of **regexp** in item 15 above, except here the return values will be vectors rather than cell arrays.

13.3   Reading user input

Sometimes, particularly in scripts (see Chapter 7), it is useful to prompt the user to input some data which are then stored in variables.

R

To read numerical data from the user, use commands such as print('Enter data:'); x = scan(). You can then enter a value, or several values separated by spaces. You can split the values up among separate lines, if you like. Press the Return/Enter key twice when you are done entering values. The values you enter will be stored in a vector.

Note that if you execute R code from an editor window, the **scan** command can behave in an undesirable way, reading its input from subsequent lines of the file rather than from the keyboard. See Section 7.6 for more information.

To read text data, you can use the command x = scan(what=character(0)). Because the default separator character used by **scan** is a space, if you enter the text "Hi there," then **x** will be a vector with two elements containing the two words. If you want to be able to enter a string containing spaces, you can use **readline** instead, which allows you to include a user prompt: x = readline('Enter string:\n'). You can also use **scan** and specify the newline as the separator character: x = scan(what=character(0), sep='\n'). This latter approach allows you to enter several strings, one per line. The result will be a vector of strings, with one vector element per line of input.

MATLAB

To read numerical data from the user, use a command like x = input('Enter data: '). Include "\n" at the end of your prompt string if you want the input to begin on a separate line. You can then enter a scalar or matrix as you would at the MATLAB command prompt. In fact, you can enter a full-blown MATLAB expression as your input. For example, you can enter **3:8**, or **sin(z)** (assuming **z** is defined), and so on. The results of evaluating the expression you type will be stored in **x**.

To read text data, you can use x = input('Enter text data: ', 's'). If you wish to read multiple lines, with one string per line, there is not a simple built-in way to do it. However, a simple loop will work. The following code reads lines, appending them to a cell array of strings, until an empty line is entered.

```
──────────────── MATLAB ────────────────
x = cell(0);  % build empty cell array
while true
    tmp = input('Enter string: ', 's');
```

```
    if (strcmp(tmp, ''))  % got empty line; stop
       break;
    end
    x = [x ; tmp];  % append new string
end
```

13.4 Recording a copy of commands and output

Both platforms offer a way to record a copy of your session's output to a file.

In R, the command sink('foo.out') will cause all subsequent output to the file **foo.out**. You will not see the output in the command window. The command split() then reverts back to normal behavior. You can use sink('foo.out', split=TRUE) to send a copy of the output to the named file, but also keep a copy visible in the command window. Note that **sink** does not also record a copy of the commands you type in the specified file; you will only see the output. Because of that, it is likely to be most useful when used from scripts.

In MATLAB, the command diary foo.out will record a copy of the commands you type, as well as their output, to the file **foo.out**. It will also continue to display them in the MATLAB command window. The command diary off reverts back to normal behavior. Note that the diary file does not contain the MATLAB command prompt ">>" next to the commands you type, so it can be more difficult to easily distinguish input from output in the file. Because of that, it is also likely to be most useful when used from scripts.

13.5 Date calculations

To demonstrate some basic calculations, consider the following tasks:

1. Calculate how many days it has been since Jan. 1, 2000.
2. Calculate how many seconds it has been since 8:30:40pm on Jan. 1, 2015.
3. Determine which date was 1,000 days after Jan. 1, 2000.
4. Calculate the time that was 3×10^7 seconds after 8:30:40pm on Jan. 1, 2000.

Note that the format strings given in the first versions of both the R and MATLAB code below are not necessary, as the formats provided are the defaults. They are included to demonstrate formatting options.

The following commands show how to perform these calculations in R:

```
──────────────────────────── R ────────────────────────────
#### Task 1
t1 = as.Date('2015-01-01', format='%Y-%m-%d')
today = as.Date(Sys.Date())
print(today - t1)
#### Task 2
t1b=as.POSIXct('2015-01-01 20:40:30', format='%Y-%m-%d %H:%M:%S')
now = as.POSIXct(Sys.time())
```

```
print(now - t1b)
#### Task 3
print(as.character(t1 + 1000, format='%Y-%m-%d'))
print(as.character(t1 + 1000, format='%d-%b-%Y'))  # default MATLAB format
#### Task 4
print(as.character(t1b + 3e7, format='%Y-%m-%d %H:%M:%S'))
print(as.character(t1b + 3e7, format='%d-%b-%Y %H:%M:%S')) # MATLAB format
```

Also see the **difftime** command for working with the differences between dates.

The following commands show how to perform these calculations in MATLAB:

```
———————————————— MATLAB ————————————————
%%% Task 1
t1 = datenum('2015-01-01','yyyy-mm-dd');
today = datenum(date);
today - t1
%%% Task 2
t1b = datenum('2015-01-01 20:40:30', 'yyyy-mm-dd HH:MM:SS');
now - t1b
%%% Task 3
datestr(t1 + 1000, 'dd-mmm-yyyy')
datestr(t1 + 1000, 'yyyy-mm-dd')  % default R format
%%% Task 4
% Note: need to convert 3e7 seconds to days
datestr(t1b + 3e7/86400, 'dd-mmm-yyyy HH:MM:SS')
datestr(t1b + 3e7/86400, 'yyyy-mm-dd HH:MM:SS')  % default R format
```

Also see the **datevec** command for working with vector representations of dates and times.

13.6   Miscellaneous

19. Pause for **x** seconds.

R	MATLAB
Sys.sleep(x)	pause(x)

20. Wait for the user to press any key.

readline() will wait until the Return/Enter key is pressed.	pause

21. Produce a beep (or a visual signal, depending on how the software and system preferences are set).

alarm()	beep

22. Display an error message and interrupt execution of the current script, function, etc.

stop('Problem!')	error('Problem!')

23. Display a warning message, and continue executing code.

`warning('Small problem!')`	`warning('Small problem!')`

24. Evaluate the contents of the string **s** as one or more commands, as if you had typed them.

`eval(parse(text=s))`	`eval(s)`

25. Measure the CPU time used to perform some commands.

`t1=proc.time()`
`    ...some commands...`
`t2=proc.time()`
`mytime = (t2-t1)[1]`
	`t1=cputime;`
	`    ...some commands...`
	`t2=cputime;`
	`mytime = t2-t1`

26. Measure the elapsed ("wall clock") time used to perform some commands.

`t1=proc.time()`
`    ...some commands...`
`t2=proc.time()`
`mytime = (t2-t1)[3]`
	`t1=clock;`
	`    ...some commands...`
	`t2=clock;`
	`mytime = etime(t2,t1)`

(MATLAB): You can also use `tic; ...commands...; mytime=toc` to measure the elapsed time. You can also save the return value of **tic** and pass it to **toc** later, to use multiple timers. For example, `myStartTime=tic; ...commands...; mytime=toc(myStartTime)`.

27. Boolean value indicating whether the value **x** is in the set **y**.

`x %in% y`	`ismember(x,y)`

If **x** is a vector, the results will be a vector of Boolean values the same length as **x**, indicating whether or not each element of **x** is in **y**.

28. The index of the location of value **x** in vector **y**.

`ind = match(x,y)`	`[tf, ind] = ismember(x,y)`

If **x** is not in the vector **y**, then R's **match** returns **NA**, while MATLAB's **ismember** returns 0. Also, if **x** is in the vector **y** multiple times, both functions return the index of the *first* element of **y** that matches **x**. (Note that in previous versions of MATLAB, the index of the *last* matching element was returned.) If **x** is a vector, then the return values of these functions are also vectors of the same size, with the information for each element of **x**. In MATLAB, to use **ismember** with strings, you should pack the strings within cell arrays, rather than regular vectors.

13.7   Debugging

I am somewhat old-school, in that my first line of action to debug misbehaving code is usually to simply add some statements to print out the values of variables that seem to be causing trouble. This can be done via some **print** statements in R, and by strategically leaving out semicolons in MATLAB code, perhaps augmented with some **sprintf** statements to make things more legible. If matrices/arrays are involved, it may be helpful to print out their dimensions, or to print out the results of calculations like `any(is.nan(v))` (R) or `any(isnan(v))` (MATLAB) to see if there are strange values somewhere in your vector.

But when printing out values is not enough, both platforms do have facilities to aid in debugging code.

R

You can insert the command `browser()` within your code (presumably a script, or a file defining a function). When that command is reached, execution is suspended, and you are able to type in R commands. The command prompt changes to **Browse[1]>** to let you know you are operating within the browser mode. For example, you can examine or modify the values of variables. You can also enter the special command **n** to execute the next line of code, or **c** to continue execution. If things get really fouled up, you can enter **Q** to terminate execution.

There is also a **debug** function that lets you tell R to enter debugging mode when a particular function is called.

MATLAB

You can insert the command `keyboard` within your code (presumably a script, or M-file defining a function). When that command is reached, execution is suspended, and you are able to type in MATLAB commands. The command prompt changes to **K>>** to let you know you are operating within keyboard mode. For example, you can examine or modify the values of variables. Enter the special command **return** to exit keyboard mode and return to normal execution. The **keyboard** command is a simple way to let you stop and check out the values of various things when your code is having trouble.

MATLAB also has a fairly standard extensive set of debugging tools for further aid in debugging. Some of the relevant commands are **dbstop**, **dbclear**, **dbstep**, and **dbcont**. You can also perform debugging via MATLAB's built-in editor. For example, if you click to the left of a line of code in the editor, a small red stop sign will appear, indicating that a breakpoint has been set at that line. When you run the code, MATLAB will suspend execution at that point. The command prompt will change to **K>>** again, to let you know you can enter debugging commands. You can use things like the "Continue" and "Step" buttons in the editor window to go through your code.

13.8   Startup and shutdown sequences

Both platforms provide mechanisms for you to specify that some code should be run when starting up and/or shutting down the software.

R

If a file **.Rprofile** exists in the current directory or in the user's home directory, its contents are sourced. It then looks for a file **.RData** in the current directory, to load a saved image (either one created when you quit R, or one saved using the methods of Section 12.9). If a function named **.First()** exists (for example, defined in your **.Rprofile**, though it could also have been saved in the .RData file), it will be called. There are many additional details and ways to customize the startup behavior, for example via environment variables in your operating system. Use the command `?Startup` for more information.

Upon shutdown, if you have defined a function named **.Last()**, it will be called.

MATLAB

If a file **startup.m** exists in the search path, its contents will be executed.

Upon shutdown, if a script file **finish.m** exists in the search path, its contents will be executed. The script can call the command **save** to save the workspace, as in Section 12.9. If for some reason you have a **finish** script which has errors in it, which is preventing you from shutting MATLAB down, you can instead use the command `quit force`. Inside your **finish** script, you can use the command `quit cancel` to abort the shutdown of MATLAB.

13.9   Add-ons: packages and toolboxes

Both R and MATLAB benefit greatly from their vast communities of users; MATLAB also of course has the advantage of being a commercial product with ongoing support and development from MathWorks.

R

One of the main sources of power in R is the vast array of packages that are available to download and install. Currently, there are more than 6,000 packages available, which perform calculations for a huge variety of particular tasks.

You can browse the list of available packages, with very short descriptions, through a CRAN mirror (see Section 1.1). You can also see a list with just the package names via the user menus in R, or via the command `install.packages()`. To actually use a package, you must do two things: first *install* the package (it is only necessary to do this once), and then *load* it (this is necessary each time you restart R and want to use the package, unless you save your workspace with the package installed). For example, to install the **deSolve** package you can use the command `install.packages('deSolve')`. You can then use `library('deSolve')` to load it. If you are writing a function in which you want to be sure the **deSolve** package is loaded, you can use `require('deSolve')`, as the **require** function returns a Boolean indicating whether the requested package is available. You can use `print(.packages())` to see a list of the names of attached (loaded) packages (you need the **print** command, because **.packages** returns its results invisibly by default; see Page 84). The command `search()` shows attached packages and data frames. To see a list of installed packages available to be attached, use `.packages(TRUE)`, or `installed.packages()` for more detailed information. You can detach a package using a command such as `detach('package:deSolve')`.

There are so many packages available, because they are user-contributed, with a fairly low barrier of entry to contribute new packages. Documentation about creating packages is available on CRAN; see Reference [24] for extensive information, or Reference [16] for a shorter and lighter overview.

MATLAB

For MATLAB, there are more than two dozen add-on products, many of them called toolboxes. Toolboxes are available for things like signal processing, statistics, wavelets, and so on. The command `ver` will show you which products you have installed. These add-ons are available from The MathWorks. An example of a non-Toolbox add-on available from The MathWorks is MATLAB Coder, mentioned at the very end of section 14.2.

There is also a MATLAB File Exchange with thousands of useful user-contributed resources available
(`http://www.mathworks.com/matlabcentral/fileexchange/`). That is just one resource available at "MATLAB Central" (`http://www.mathworks.com/matlabcentral/`); there are also interesting challenges called Cody problems, plots, and questions and answers about MATLAB.

13.10   Object-oriented programming

Both platforms have facilities for object-oriented programming. In R, there are two different approaches, or types of objects: S3 and S4. It is somewhat easier to write S3 objects, by providing a set of functions with particular special names that can operate on your objects. S4 objects are more formal, with more mechanisms to catch errors or abuses, but with a bigger learning curve to go along with them. See Reference [16] for a very brief introduction to writing S3 objects, and Reference [26] for more information.

In MATLAB, you can provide a collection of functions all within a folder named after the object class you are implementing. It is also possible to write a single file containing the class definition and methods for a given class. See the various resources linked from `http://www.mathworks.com/discovery/object-oriented-programming.html` for more information.

13.11   Other interfaces

We are well past the days where everyone does all of their computations sitting at a desktop or laptop computer. Many different devices, capable of varying levels of computations, now pervade our lives. Both R and MATLAB have support for additional devices. There are also additional software interfaces available.

R

One very popular third-party interface for R is called **RStudio**. It is available in a free, open-source edition as well as a commercial version from `http://www.rstudio.com`. RStudio provides a richer integrated development environment, with debugging tools, an editor that performs syntax-aware text coloring, a workspace inspector showing your defined variables, and so on. It is widely used among the students in my classes because of its superior editor. It also provides facilities to run R on a server and access it via the Web.

Other third-party interfaces for R are R Commander, available for download from `http://www.rcommander.com`, and Tinn-R, available from `http://nbcgib.uesc.br/lec/software/editores/tinn-r/en`. There is also an interface to R available for the Emacs editor, called **ESS** ("Emacs Speaks Statistics"), available from `http://ess.r-project.org`.

There is a tool called **Sweave** which allows you to embed R commands within a LaTeX file to aid in document production incorporating R commands and output, available at `http://www.stat.uni-muenchen.de/~leisch/Sweave`. This is similar to MATLAB's **publish** feature.

It is also possible to install and run R directly on a Raspberry Pi device.

MATLAB

Because the standard MATLAB user interface is so well developed, there has not been a need for third parties to provide alternative interfaces. However, there is a MATLAB mode for the Emacs text editor available (`http://matlab-emacs.sourceforge.net`), which aids in editing MATLAB files (scripts and functions), and also allows you to run MATLAB commands from within a shell inside of Emacs.

The **publish** command lets you export MATLAB code and its resulting output in a variety of formats, including HTML, PDF, LaTeX, and so on. Like the Sweave tool for R, this is useful for producing documents based on MATLAB code.

It is possible to run MATLAB from mobile devices, including iOS devices such as iPhone and iPad and Android devices (`http://www.mathworks.com/mobile`). This is done by either accessing an instance of MATLAB running on your own computer, or through The MathWorks Cloud. The MATLAB mobile app for Android can aquire data from built-in sensors on Android devices, including their accelerometer, orientation, and position sensors.

MATLAB can aquire data from sensors attached to a Raspberry Pi via the MATLAB Support Package for Raspberry Pi Hardware available from The MathWorks. At this time, MATLAB itself does not run as a standalone program on a Raspberry Pi.

13.12   Efficiency/performance

Because both R and MATLAB are interpreted languages, the same basic primary advice applies to both platforms: *avoid writing explicit **"for"** loops if vectorized statements can be used instead.* That is, rather than writing this in R:

────────────────────────── R ──────────────────────────
```
mysum = 0
for (i in 1:1000000)
  mysum = mysum + i
```

it is better to just write `mysum = sum(1:1000000)`.[1] Not only is the latter code shorter, it also runs much more quickly, because the implied **for** loop which still takes place runs internally at compiled speeds, rather than explicitly at interpreted speeds.

The MathWorks has been getting fairly aggressive at making performance enhancements in MATLAB over the past several years. MATLAB will now take small chunks of iterative code that are relatively uncomplicated, and precompile them so that they do not need to be reinterpreted on every iteration. As a result, the following explicit MATLAB code:

────────────────────────── MATLAB ──────────────────────────
```
mysum=0;
for i=1:1000000
```

---

[1]Technically, the above code will suffer integer overflow, because when you write **a:b**, the values are stored as integers. You should instead use `as.numeric(1:1000000)` to force R to store the values as double-precision floating point numbers to avoid overflow.

```
    mysum = mysum+i;
end
```

runs more than 100 times faster than the equivalent R code above, on my reference system. MATLAB will optimize the code within a script or function, as long as it is not deemed to be too complex. Breaking larger computations down into smaller functions and scripts may improve performance.

The second major piece of advice also applies equally to both platforms: *avoid repeatedly growing a matrix or vector; preallocate additional space if you think you will need it.* An example of inefficient code would be the following MATLAB commands:

```
————————————————————— MATLAB —————————————————————
x = 1;
for i = 2:1000
  x(i) = i^2;
end
```

This begins with a 1×1 scalar **x**, and then repeatedly adds a new element to the end; when the code finishes, **x** is a 1×1000 vector. The problem is that every time a new element is added to the end of the vector, and new block of memory must be allocated for the new, larger vector, and then the previous contents copied over to the new area in memory. The same problem exists in R. It is better to preallocate space for the vector (for the MATLAB code above, placing the line x = zeros(1,1000); at the beginning will allocate the needed memory for the vector).

As with most computing environments, drawing graphics and doing input/output to a file or the command console is usually at least an order of magnitude slower than performing raw computations. If you can save the graphics and file I/O until after computations are finished, your code may run much more quickly.

For discussion of some issues in R which include performance, see the somewhat entertaining Reference [3]. For more information about performance issues in MATLAB, see the comprehensive Reference [1].

14

Calling C

Both R and MATLAB can perform vectorized computations, where iteration over the elements of a vector or matrix is performed implicitly, i.e., internally. This is much more efficient than writing explicit "**for**" loops, which run very slowly due to both platforms being interpreted (rather than compiled) languages. For some computations, however, explicit iteration is required. This occurs when each stage of a computation depends on the results of the previous stage. The following two examples of such computations will be used for demonstration purposes here.

1. The Fisher Yates shuffle [8] for permuting a set of n values in a vector $\mathbf{v}$; here we consider a more recent version, sometimes known as the Knuth shuffle [6, 15], which is more efficient. The algorithm is as follows:

    Let $\mathbf{i}$ count from $\mathbf{n}$ down to 2
        Let $\mathbf{j}$ be a random integer between 1 and $\mathbf{i}$, inclusive
        Exchange the elements in positions $\mathbf{i}$ and $\mathbf{j}$ of $\mathbf{v}$

2. A stochastic birth-death process simulated for s events. In a given population at time t, each individual gives birth at rate ϕ, and dies at rate μ, according to a Poisson process. We build two vectors, one containing the time of each event (birth or death), and the other containing the population sizes after the corresponding events at those times. Let n be the population size at the current time. The simulation proceeds as follows:

    Let $\mathbf{i}$ count from 1 to $\mathbf{s}$
        Generate an exponentially distributed random value with mean $1/(n(\phi + \mu))$
        Add the above value to the current time, as the time of the next event
        With probability $\phi/(\phi + \mu)$, the event is a birth (and n is incremented),
            otherwise the event is a death (and n is decremented)
        Append the new event time and population size to vectors

Note that the C functions in this chapter call a random-number generator. The details of such calls are generally very specific to the operating system and/or C compiler you are using. With the gcc compiler under Mac OS-X and Linux, the function **random()** returns a random integer stored as a "long int," while the function **srandom(seed)** can be called with an unsigned int to reset the random number generator's seed. In other C compilers, there are sometimes functions **rand** and **srand** which behave in an equivalent way (though on some systems, **rand** is a very poor-quality generator, in the sense that the values it produces may have cyclic patterns in some of their bits). You can also search for C code to generate pseudorandom numbers; for example the "Mersenne twister" is a popular generator for which source code is freely available.

14.1   R

14.1.1   Example and overview

A R-compatible C function implementing the shuffle is given below. We would like the first argument to the function to be a vector of double-precision values to sort, and the second argument to be an integer saying how long the vector is.

```
──────────── Cshuffle.c ────────────
#include <stdlib.h>  /* for declaration of random() */
#include <math.h>    /* for declaration of floor() */

/* Permute the elements of a vector using the Fisher-Yates shuffle */
void Cshuffle(double *v, int *nptr)
{
  int i, j, n = nptr[0];
  double tmpDbl;

  for (i=n; i > 1; i--) {
    /* Generate a random integer from 0 to i-1
     * (not from 1 to i, because vectors in C use 0-based indexing). */
    j = (int)floor((((double) random())) / (RAND_MAX + 1.0) * i);
    /* swap v[j] with the last element of v */
    tmpDbl = v[j];  v[j] = v[i-1];  v[i-1] = tmpDbl;
  }
}
```

Assume the function above is stored in a file called **Cshuffle.c**. To compile this function into a form usable by R, you should enter the following command in a terminal window (not at the R command prompt), in the same directory or folder containing the C file:

```
R CMD SHLIB Cshuffle.c
```

On Unix systems (Mac or Linux) this will most likely produce a file called **Cshuffle.so**; it should produce **Cshuffle.dll** on Windows. If you do not see the appropriate file, look around and see what suffix was used. (Ignore the **Cshuffle.o** file; it is just the compiled object code.)

The function can then be read into R by using the following command:

```
──────────── R ────────────
dyn.load('Cshuffle.so')  # use 'Cshuffle.dll' in Windows
```

The function cannot simply be called like any built-in or user-defined function in R, i.e., you cannot simply say `Cshuffle(v,n)`. Instead, it must be called via the special routine `.C`. For example, to test out the function by permuting the values from 1 to 20, the following commands can be used in R:

```
──────────── R ────────────
v = as.double(1:20)
n = length(v)
tmp = .C('Cshuffle',v,n)
permutedv = tmp[[1]]
```

There are several things to note about the above process.

1. R passes scalar values to C functions inside vectors of length 1. This is why the second parameter in the C function is declared as **int *nptr** rather than simply **int n**. Inside our function, we extract the (single) element of this vector into the scalar C variable **n** for clarity.

2. Care must be taken to ensure that data types are correct. For example, the R command v = 1:20 produces a vector of integers, rather than doubles; v = as.double(1:20) ensures that the vector contains double-precision floating-point values. Similarly, n = length(v) produces an integer as needed; if we had simply written n = 20 instead, it would have been a double. (However, n = as.integer(20) would work.) Great care must be taken here, as if the data types are incorrect, you likely will not see any explicit warnings or errors — you may instead simply get incorrect results (see the **testprint** example on Page 182).

3. Matrices are passed in to C functions as vectors with column-major order (i.e., values arranged column by column), so passing matrix **A** to a C function is equivalent to passing the vector **c(A)**.

4. The C function cannot use **return** to pass return values to the caller, but instead should be declared to have type **void**. Values must be returned to the caller by modifying the vectors passed in as arguments, i.e., the function should have "side effects" on the vectors. This means if you want to write a function which takes no arguments in but which returns a vector, you must in fact pass in a vector of the appropriate length.[1]

5. When the C function is done, the original R values passed as arguments are not actually modified. Instead, copies of those values are passed to the .**C()** function, and a list is returned containing the modified vectors. That is why the command permutedv = tmp[[1]] was used to extract the results above; the first item in the returned list was the modified vector **v**, and the second item in the list was the (unmodified) value **n**.

6. The name of the function does not need to match the name of the C file, although it is easier to keep track of things if it does. When you call the function using .**C()**, use the name of the actual function, rather than the name of the file.

7. If there are multiple functions in your C file, they will all be loaded (and callable from R) when you load your .**so** or .**dll** file via **dyn.load()**.

8. If you load two different files, say **test1.so** and **test2.so** which both define functions named **testfunc**, then calling .**C('testfunc', ...)** will call whichever was loaded more recently. You can specify which version to use via an extra PACKAGE argument (omit the filename's suffix), e.g., .**C('testfunc', ..., PACKAGE='test1')**.

9. When you use the terminal command **R CMD SHLIB foo.c** to compile your C code, be aware of whether the command **R** in the terminal is running the 32-bit or 64-bit version of R. When trying to run your code, you must ensure that the interactive version of R you are using has the same architecture as the version used for compiling. Otherwise, you will see an error about being unable to load the shared object; it will likely complain that the file has the wrong architecture. You can use the Terminal command R -q -e 'version$arch' to try and determine if it the command-line version of R is the 32-bit version (you

---

[1] I originally considered writing the **Cshuffle** function to take a single integer argument **n** and return a vector of the permuted values from 1 through **n**, but because of the above constraints, it was no more difficult to write the function to receive a vector containing general values to permute.

will likely see something like "i386") or the 64-bit version (you will see "x86_64"). The details are dependent on the processor architecture, operating system, and version of R. Entering the command `version$arch` at the command prompt within R will give the same information for the interactive version of R you are running.

10. If your C file "mycode.c" contains a function "**void R_init_mycode(DllInfo *info)**", it will be called when your file is loaded via **dyn.load**. This gives you an opportunity to perform any initializations needed for your code to run. Similarly, if the function **void R_unload_mycode(DllInfo *info)** exists, it will be called if you unload your code via the **dyn.unload** R function.

Because of items 2–4 above, it is usually easier (and safer) to write an R wrapper around the C function you wish to call, to ensure that the data types are correct and to extract the results. For example, the following R function makes it easier to call the C **Cshuffle** function. You simply give it a vector **v** (which may contain integer or double values), and it returns the permuted vector.

```
shuffle.R
shuffle = function(v) {
  if (!is.numeric(v))
    stop('Cshuffle(): Non-numeric v provided!')
  tmp = .C('Cshuffle', as.double(v), length(v))
  return(tmp[[1]])
}
```

Note that using a wrapper can make things run less efficiently, depending on how much data manipulation the wrapper performs. A leaner wrapper function was also written as below, which does not check the data-types of the vector **v**:

```
shuffle2.R
shuffle2 = function(v) {
  tmp = .C('Cshuffle', v, length(v))
  return(tmp[[1]])
}
```

Code similar to that shown below was used to measure the time needed to repeatedly permute a set of values in six different ways:

1. Using **.C()** to call the **Cshuffle** C function directly.
2. Using the R wrapper function **shuffle2**.
3. Using the R wrapper function **shuffle**.
4. Using the **sample** function.
5. Using **.Call()** with the **Cshufflecall** C function (see Page 186).
6. Using **.C()** to call the **Cshuffle2** C function, which uses R's random-number generator (see Page 183).

As shown in Table 14.1, when permuting a large vector (of length $n = 2 \times 10^5$) 1,000 times, using **.C()** ran in about 74% of the time it took **sample()** to permute the same values.

```
R
iters = 1000;  v=as.double(1:200000);  n=length(v)
t1=proc.time()
for (i in 1:iters) {
```

Method	Raw times			Relative to **sample**		
	User	System	Elapsed	User	System	Elapsed
.C('Cshuffle'):	6.84	0.174	6.96	0.801	1.6	0.812
shuffle2:	8.23	0.142	8.3	0.964	1.3	0.968
shuffle:	8.24	0.144	8.32	0.965	1.32	0.97
.C('Cshuffle2'):	8.07	0.139	8.14	0.945	1.28	0.95
sample:	8.54	0.109	8.58	1	1	1
.Call('Cshufflecall'):	7.02	0.07	7.03	0.822	0.642	0.82

TABLE 14.1
Timing results (in seconds) of permuting the integers from 1 to 2×10^5, repeated 1000 times, using six different methods. Raw times are given in seconds; the times relative to using **sample** are also shown, indicating that the C code can run in as little as three-quarters the time of R's **sample** function for large vectors.

```
# uncomment one of the following lines to time one of the methods
tmp = .C('Cshuffle',v,n)[[2]]
# tmp = shuffle2(v)
# tmp = shuffle(v)
# tmp = .C('Cshuffle2',v,n)[[2]]
# tmp = sample(v)
# tmp = .Call('Cshufflecall', v, n)
}
t2=proc.time();  print(t2-t1)
```

14.1.2   Printing, warnings, and errors

If you use the **printf()** function in your C code, you may not see the output in the R console.[2] The supported way to print values from C code is to use the function **Rprintf()**, which behaves like **printf()** but integrates its output into R's output stream. Just include the file **R.h** first, as in the example below. There is also a command **REprintf()** for printing on the **stderr** error stream.[3]

To display a warning, you can call the function **warning**; to display an error (the equivalent of using the R function **stop**), call the C function **error**. Both of these functions accept the same types of arguments as **printf**, i.e., either a simple string, or a formatting string followed by some parameters. In the example below, a warning is produced if the first parameter is odd, and if it is negative, an error is generated.

```
──────────── testprint.c ────────────
#include <R.h>

void
testprint(int *nptr)
{
  int n;
```

---

[2] In Mac OS-X 10.6.8 with R version 2.14.1, output from **printf()** is shown, but in a faint gray font which is not very easy to see.

[3] These messages may be rendered differently, for example in a red font, depending on your operating system.

```
  n = nptr[0];
  if (n % 2)
    warning("n was odd; it had value %d\n", n);
  if (n < 0)
    error("n was negative!  It had value %d\n", n);
  Rprintf("n=%d\n", n);
  REprintf("Now on the error stream: n=%d\n", n);
}


void
R_init_testprint()
{
  Rprintf("hi there, initializing testprint code!\n");
}
```

Note that calling this function via **.C('testprint', 3)** shows that the function believes **n** is 0, because the value 3 is passed in with type "double" rather than "integer." Calling **.C('testprint', as.integer(3))** gives the correct behavior.

14.1.3   Random numbers

The **Cshuffle.c** code on Page 178 calls the C library function **random()** to produce a random integer from 0 to **RAND_MAX**. It also therefore uses the C library's random number seed and state; on Unix systems, this will produce the same sequence of random numbers every time you restart R and call the code. You may wish to seed C's random number generator using the current time; on Unix systems, this can be done by including the following function in the **Cshuffle.c** file, and relying on R's mechanism for initializing the dynamically loaded file:

```
──────────────────── R ────────────────────
/** Seed the random number generator, based on the current time **/
void
R_init_Cshuffle(DllInfo *info)
{
    struct timeval tp;  struct timezone *tzp;

    tzp = NULL;  gettimeofday(&tp, tzp);
    srandom((unsigned int) tp.tv_usec);
}
```

An alternative would be to have the **Cshuffle** function accept an additional parameter which, if it is not zero, gets used as a random number seed.

The advantage of using the C library's random number generator is that, other than the interface function that you call from R, your C code will be more portable in the sense that you can use it as part of a stand-alone C program. A disadvantage is that the built-in random number generator may not be very good, depending on what compiler you are using.

Another option is to call R's random number generators. Three basic generators are provided which are callable directly from C; they are shown in Table 14.2. If you use any of R's random number routines from C, note that you must also call GetRNGstate()

Function	Description
`double unif_rand()`	Generate a single random value from the continuous uniform distribution from 0 to 1. The value 0 may be returned; the value 1 should never be returned.
`double norm_rand()`	Generate a single random value from the standard normal distribution $N(0,1)$.
`double exp_rand()`	Generate a single random value from the exponential distribution with mean 1.

TABLE 14.2
Basic functions which may be called from C code to produce random values. Placing the statement `#include <R.h>` at the top of your C code will include the declarations of these functions.

before calling them, and `PutRNGstate()` afterward. These ensure that your calls to the C random variate generators are integrated into the sequence of states of R's random number generator. Below is a version of the **Cshuffle.c** code which uses `unif_rnd()`.

```
_____ Cshuffle2.c _____
#include <R.h>     /* for declaration of random-number routines */
#include <math.h>    /* for declaration of floor() */

/* Permute the elements of a vector using the Fisher-Yates shuffle
 * using R's random number generator */
void Cshuffle2(double *v, int *nptr)
{
  int i, j, n = nptr[0];
  double tmpDbl;

  GetRNGstate();
  for (i=n; i > 1; i--) {
    /* Generate a random integer from 0 to i-1
     * (not from 1 to i, because vectors in C use 0-based indexing). */
    j = (int)floor(unif_rand() * i);
    /* swap v[j] with the last element of v */
    tmpDbl = v[j];  v[j] = v[i-1];  v[i-1] = tmpDbl;
  }
  PutRNGstate();
}
```

You may find it convenient to call the internal C versions of the various R routines related to random numbers. Some examples of such functions are listed below. You should place the statement `#include <Rmath.h>` at the top of your C code to include declarations of the functions below.

`double rexp(double m)`: Generate a single random value from the exponential distribution with mean m (note that this is in contrast with the R function **rexp**, where you provide the rate, i.e., the reciprocal of the mean, and can also specify how many random values you need).

double dexp(double x, double m, int log): Analogous to R function, except the mean m should be given, rather than its reciprocal.

double pexp(double q, double m, int lower.tail, int log.p): Analogous to R function, except the mean m should be given, rather than its reciprocal.

double qexp(double p, double m, int lower.tail, int log.p): Analogous to R function, except the mean m should be given, rather than its reciprocal.

See the file **Rmath.h** for declarations of the many other similar C counterparts of R random number functions, and to check the argument types. For example, there is a function **double rbinom(double size, double prob)**, but note that its **size** parameter and its return type are both **double** (rather than **int** as one might expect). You must also call GetRNGstate() and PutRNGstate() before and after calls to the various functions such as rexp, rbinom, and so on, to integrate them into the sequence of R's random number stream.

Below is a C function implementing a stochastic birth-death process, which calls the exponential random number generator **rexp** and the uniform generator **unif_rand**. Equivalent R code follows. For this simulation, the R version takes roughly 160 times as long to run as the C version.

```
―――――――――――――――――――――― Cbirthdeath.c ――――――――――――――――――――――
#include <R.h>   /* decl's of unif_rand(), GetRNGstate(), PutRNGstate() */
#include <Rmath.h>  /* for declaration of rexp() */


/* Simulate stochastic birth-death process with per-capita
 * birth rate phi and per-capita death rate mu.
 * eventTimes and popVec must be vectors of length s,
 * large enough to hold data for all of the time steps.
 */
void Cbirthdeath(double *phiptr, double *muptr,int *sptr,
                 double *eventTimes, int *popVec)
{
  int i, s = sptr[0];
  double phi = phiptr[0], mu = muptr[0], birthProb, r;

  birthProb = phi/(phi+mu);  /* probability a given event is a birth */
  eventTimes[0] = 0.0;  popVec[0] = 100;
  GetRNGstate();
  for (i=0; i < s-1; i++) {
    /* Inter-event time: generate a random value from exponential
     * distribution with mean 1/(popSize*(phi+mu)) */
    r = rexp(1.0/(popVec[i]*(phi+mu)));
    eventTimes[i+1] = eventTimes[i] + r;
    if (unif_rand() < birthProb)
      popVec[i+1] = popVec[i] + 1;  /* event is a birth */
    else
      popVec[i+1] = popVec[i] - 1;  /* event is a death */
  }
  PutRNGstate();
}
```

```
―――――――――――――――――――――― birthdeath.R ――――――――――――――――――――――
birthdeath = function(phi, mu, s) {
  birthProb = phi/(phi+mu)  # probability a given event is a birth
```

	Raw times			Times faster than R code		
Method	User	System	Elapsed	User	System	Elapsed
.C('Cbirthdeath')	0.288	0.031	0.316	175	59.6	164
.C('Cbirthdeath2')	0.258	0.06	0.315	195	30.8	164
birthdeath.R	50.3	1.85	51.7	1	1	1
.External('Cbirthdeathexternal')	0.207	0.011	0.216	243	168	240

TABLE 14.3
Timing results (in seconds) of simulating the birth-death process for $s = 3 \times 10^6$ events with $\phi = 1.1$ and $\mu = 1$ using four different methods. The C code generally runs roughly 200 times as quickly as the R version. Note that among different replicates of the experiment, different C versions ran in the fastest time, i.e., these timings do not indicate that the **.External** version of the C code is necessarily the fastest.

```
  eventTimes = double(s);  popVec = integer(s)
  eventTimes[1] = 0.0;  popVec[1] = 100
  for (i in 1:(s-1)) {
    # Inter-event time: generate a random value from exponential
    # distribution with mean 1/(popSize*(phi+mu))
    r = rexp(1,popVec[i]*(phi+mu))
    eventTimes[i+1] = eventTimes[i] + r
    if (runif(1) < birthProb) {
      popVec[i+1] = popVec[i] + 1  # event is a birth
    } else {
      popVec[i+1] = popVec[i] - 1  # event is a death
    }
  }
  return(list(eventTimes,popVec))
}
```

An additional C version was used, **Cbirthdeath2.c**, which used the C library's random number generator rather than R's, to produce the two random values needed per event. This was done by changing the two relevant lines in the code to:

```
———————————————————— R ————————————————————
r = -log((random()+1.0)/(RAND_MAX+1.0))/(popVec[i]*(phi+mu));
if (random()/(double)RAND_MAX < birthProb)
```

Timings of the C and R versions of the code, along with another version using the **.External** method for invoking code (see Page 187) are shown in Table 14.3. The C code runs approximately 200 times faster than the R code, which is typical for sequential simulations such as this.

14.1.4 More advanced features

If you wish to work more closely with native R variables, pass lists to your function, or pass a variable number of arguments to your function, the **.C** mechanism is not the best way to proceed. Instead, there are two other ways of calling C code from R: **.Call** and **.External**. They require more complex C code to handle function arguments, set up return values, and handle memory management. Their facilities for dealing with function arguments is somewhat similar to MATLAB's mechanism, requiring you to extract data from generic data pointers. The primary difference between **.Call** and **.External** is in how they pass arguments to your C function; the latter is more appropriate if you want to allow a variable number of arguments, i.e., optional parameters.

C variables of type SEXP ("Simple EXPression") are used to refer to arguments to your function. Facilities are provided for obtaining pointers to the actual data within those arguments, for determining and coercing the data types, and so on. Because R uses garbage collection to periodically clean up allocated memory which is not being used, you must request that R protect the memory used by newly allocated variables within your code. This can be done via the PROTECT function. When your code is about to return, you can instruct R to UNPROTECT the last several items that were protected. See the **Cshuf-flecall.c** code below for an example, which can be called via .Call('Cshufflecall', as.double(1:10), as.integer(10)).

```
————————————————— Cshufflecall.c —————————————————
#include <R.h>    /* for declaration of random-number routines */
#include <Rinternals.h>  /* for dealing with SEXP stuff */
#include <math.h>    /* for declaration of floor() */

/* Permute the elements of a vector using the Fisher-Yates shuffle */
SEXP Cshufflecall(SEXP Rv, SEXP Rn)
{
  R_len_t i, j, n;
  double *v, *newv, tmpDbl;
  SEXP Rnewv;

  if (!isReal(Rv)) error("Vector v did not have real/double values");
  if (!isInteger(Rn)) error("n was not of type integer");
  v = REAL(Rv);  /* extract pointer to actual data for v */
  n = INTEGER(Rn)[0];  /* extract scalar value of n */
  if (n > length(Rv))  /* check n against length of vector v */
    error("Cshufflecall: n was %d, but vector v only had length %d\n",
          n,length(Rv));
  /* allocate memory for return value, tell R not to garbage-collect it */
  PROTECT(Rnewv=allocVector(REALSXP, n));
  newv = REAL(Rnewv);  /* extract pointer to actual data for newv */

  GetRNGstate();
  for (i=0; i < n; i++)  /* copy original v to newv to work on it */
    newv[i] = v[i];
  for (i=n; i > 1; i--) {
    /* Generate a random integer from 0 to i-1
     * (not from 1 to i, because vectors in C use 0-based indexing). */
    j = (int)floor(unif_rand() * i);
    /* swap newv[j] with the last element of newv */
    tmpDbl = newv[j];  newv[j] = newv[i-1];  newv[i-1] = tmpDbl;
  }
  PutRNGstate();
  UNPROTECT(1);  /* we don't need to protect newv any more */
  return(Rnewv);
}
```

For an example of using **.External** to allow for a variable number of parameters, see the following version of the birth-death simulation. It returns a list containing two vectors: a vector of real numbers with the event times, and a vector of integers with the corresponding population sizes. It assumes default values for its three parameters **phi**, **mu**, and **s**. These can be overridden by providing named arguments. That is, the simulation can be called via tmplist = .External('CbirthdeathExternal'), or tmplist = .External('CbirthdeathExternal', s=as.integer(50), phi=1.4), and so on. The function does not take unnamed arguments, as it would require lengthier C code to handle them as well.

```
―――――――――――――――― Cbirthdeathexternal.c ――――――――――――――――
#include <R.h>  /* decl's of unif_rand(), GetRNGstate(), PutRNGstate() */
#include <Rmath.h>  /* for declaration of rexp() */
#include <Rinternals.h>  /* for dealing with SEXP stuff */

/* Simulate stochastic birth-death process with per-capita
 * birth rate phi and per-capita death rate mu.
 * Return a list containing two vectors: eventTimes and popVec
 */
SEXP CbirthdeathExternal(SEXP args)
{
  int i, s=1000, *popVec;
  double phi=1.1, mu=1.0, birthProb, r, *eventTimes;
  SEXP returnList, eventTimesSXP, popVecSXP;

  args = CDR(args);  /* remove function name from args */
  Rprintf("length(args) is now %d\n", length(args));
  for (i=0; args != R_NilValue; i++, args = CDR(args)) {
    if (isNull(TAG(args)))
      error("Cbirthdeath(): unnamed arguments not allowed");
    /* extract name of next argument */
    const char *name = CHAR(PRINTNAME(TAG(args)));
    SEXP el = CAR(args);  /* and the value of that argument */
    if (length(el) != 1)
      error("Cbirthdeath(): All arguments must be scalars");
    if (!strcmp(name, "phi")) {
      if (TYPEOF(el) != REALSXP)
        error("Cbirthdeath(): phi must be a real value");
      phi = REAL(el)[0];
    } else if (!strcmp(name, "mu")) {
      if (TYPEOF(el) != REALSXP)
        error("Cbirthdeath(): mu must be a real value");
      mu = REAL(el)[0];
    } else if (!strcmp(name, "s")) {
      if (TYPEOF(el) != INTSXP)
        error("Cbirthdeath(): s must be an integer value");
      s = INTEGER(el)[0];
    } else {
      error("Cbirthdeath(): unrecognized argument name `%s'", name);
    }
  }
  Rprintf("phi=%g, mu=%g, s=%d\n", phi, mu, s);
```

```
/* allocate memory for the two return vectors */
PROTECT(eventTimesSXP=allocVector(REALSXP, s));
PROTECT(popVecSXP=allocVector(INTSXP, s));
/* then allocate memory for the list of length 2 */
PROTECT(returnList=allocVector(VECSXP, 2));
/* put the two vectors into the list */
SET_VECTOR_ELT(returnList, 0, eventTimesSXP);
SET_VECTOR_ELT(returnList, 1, popVecSXP);
/* get pointers to the two vectors' actual data */
eventTimes = REAL(eventTimesSXP);
popVec = INTEGER(popVecSXP);

birthProb = phi/(phi+mu);  /* probability a given event is a birth */
eventTimes[0] = 0.0;  popVec[0] = 100;
GetRNGstate();
for (i=0; i < s-1; i++) {
  /* Inter-event time: generate a random value from exponential
   * distribution with mean 1/(popSize*(phi+mu)) */
  r = rexp(1.0/(popVec[i]*(phi+mu)));
  eventTimes[i+1] = eventTimes[i] + r;
  if (unif_rand() < birthProb)
    popVec[i+1] = popVec[i] + 1;  /* event is a birth */
  else
    popVec[i+1] = popVec[i] - 1;  /* event is a death */
}
PutRNGstate();
UNPROTECT(3);  /* no more need to protect the 3 things we allocated */
return(returnList);
}
```

For more information about **.C**, **.Call**, **.External**, and interfacing with C and other languages in general, consult the comprehensive reference [24].

14.2   MATLAB

14.2.1   Example and overview

A MATLAB-compatible C function to perform the shuffle is below, along with an additional mandatory function called a gateway function. The gateway function checks and manages arguments to the C function, and allocates memory for return values. In many ways, this is similar to R's **.Call** mechanism for interfacing with C functions, as arguments to the function are encapsulated within special data structures defined by MATLAB. See Reference [13] for more information.

```
————————————————— Cshuffle.c —————————————————
#include "mex.h"
#include <stdlib.h>  /* for declaration of random() */
#include <math.h>    /* for declaration of floor() */
```

```c
/* Permute the elements of a vector using the Fisher-Yates shuffle */
void Cshuffle(double *outv, double *v, int n)
{
  int i, j;
  double tmpDbl;

  for (i=0; i < n; i++)
    outv[i] = v[i];
  for (i=n; i > 1; i--) {
    /* Generate a random integer from 0 to i-1
     * (not from 1 to i, because vectors in C use 0-based indexing). */
    j = (int)floor((((double) random())) / (RAND_MAX + 1.0) * i);
    /* swap outv[j] with the last element of outv */
    tmpDbl = outv[j];  outv[j] = outv[i-1];  outv[i-1] = tmpDbl;
  }
}



/* The gateway function */
void mexFunction( int nlhs, mxArray *plhs[],
                  int nrhs, const mxArray *prhs[])
{
  int n;
  unsigned int rseed;
  double *v, *outv;

  /* Check for proper number of arguments */
  if ((nrhs < 1) || (nrhs > 3))
    mexErrMsgIdAndTxt("RMatlab:Cshuffle:nrhs",
                      "One--three arguments needed: vector n randseed");
  if (nrhs == 1)
    /* if length of vector wasn't specified, use entire vector */
    n = mxGetNumberOfElements(prhs[0]);
  else {
    /* Make sure second argument is a real scalar */
    if (!mxIsDouble(prhs[1]) || mxIsComplex(prhs[1]) ||
        mxGetNumberOfElements(prhs[1]) != 1)
      mexErrMsgIdAndTxt("RMatlab:Cshuffle:invalidArgument",
                        "Second argument must be a real scalar");
    /* Get length of vector */
    n = (int)mxGetScalar(prhs[1]);
    if (nrhs == 3) {
      /* Make sure second argument is a real scalar */
      if (!mxIsDouble(prhs[2]) || mxIsComplex(prhs[2]) ||
          mxGetNumberOfElements(prhs[2]) != 1)
        mexErrMsgIdAndTxt("RMatlab:Cshuffle:invalidArgument",
                          "Third argument must be a real scalar");
      rseed = (unsigned int)mxGetScalar(prhs[2]);
      srandom(rseed);
    }
  }
```

```
   if (nlhs > 1)
     mexErrMsgIdAndTxt("RMatlab:Cshuffle:nlhs",
                        "Only one return value needed");

   /* Make sure first argument is a vector, not a matrix */
   if (mxGetM(prhs[0]) != 1 && mxGetN(prhs[0]) != 1)
     mexErrMsgIdAndTxt("RMatlab:Cshuffle:1stArgNotVector",
                        "First argument must be a vector");

   /* Create pointer to data in input vector */
   v = mxGetPr(prhs[0]);

   /* Create row vector for return argument */
   plhs[0] = mxCreateDoubleMatrix(1,n,mxREAL);
   outv = mxGetPr(plhs[0]);

   Cshuffle(outv,v,n);
}
```

The file can be compiled by entering the following command at the MATLAB prompt (just be sure the working directory is the one containing the C file):

———————————————————— MATLAB ————————————————————
```
mex Cshuffle.c
```

This will produce a file with a new suffix which depends on your operating system and processor architecture; on my machine, it produces a file named **Cshuffle.mexmaci64**.

Your function can then be called like any other MATLAB function. For example, to permute the values from 1 to 20, you can simply do the following:

———————————————————— MATLAB ————————————————————
```
v = 1:20; n = length(v);
permutedv = Cshuffle(v,n);
```

Code similar to that shown below was used to measure the time needed to repeatedly permute a set of values in three different ways:

1. **Cshuffle(v,n)**.

2. **Cshuffle2(v,n)** (see Page 194).

3. The MATLAB command **randperm(n)**.

As shown in Table 14.4, when permuting the integers from 1 to 2×10^5 1,000 times, the **Cshuffle** code runs in approximately 14% of the time (User CPU time) compared with performing the equivalent task via randperm(n).

———————————————————— MATLAB ————————————————————
```
iters = 1000; v=1:200000; n=length(v);
t1=cputime;
tic
for i = 1:iters
  % uncomment one of the following lines to time one of the methods
  tmp = Cshuffle(v,n);
  % tmp = Cshuffle2(v,n);
```

	Raw times		Relative to **randperm**	
Method	User	Elapsed	User	Elapsed
Cshuffle:	3.65	3.64	0.139	0.221
Cshuffle2:	554	553	21.1	33.5
randperm:	26.2	16.5	1	1

TABLE 14.4
Timing results (in seconds) of permuting the integers from 1 to 2×10^5, repeated 1,000 times, using three different methods. Raw times are given in seconds; the times relative to using **randperm** are also shown, indicating that the C code can run in as little as 14% the time of MATLAB's **randperm** function for large vectors.

```
  % tmp = randperm(n);
end
toc
t2=cputime; disp(sprintf('CPU time = %5.3g', t2-t1))
```

Things to note here are:

1. The main MATLAB data structure available to your C code is the **mxArray**. The pointer to the actual data within an **mxArray** variable **m** can be obtained via **dblPtr = mxGetPr(&m)**, where **dblPtr** is of type **double ***.

2. Your gateway function receives four parameters: (1) **nlhs**, the Number of Left-Hand Sides (i.e., the number of return values); (2) **plhs**, a vector of pointers to **mxArray** items where the return values can be stored; (3) **nrhs**, the Number of Right-Hand Sides (i.e., the number of arguments provided to your function); and (4) **prhs**, a vector of pointers to **mxArray** items containing those arguments.

3. The values in **prhs** should be considered read-only. Do not modify them, as doing so may have bad results.

4. The actual values within a matrix in an **mxArray** are stored in column-by-column order. That is, for a 3×3 matrix, **dblPtr[0]** refers to the element in row 1, column 1, **dblPtr[3]** refers to the element in row 1 column 2, and so on.

5. You must allocate memory for any values you wish to return, typically with **mxCreateDoubleMatrix(m, n, mxREAL)** to create an $m \times n$ matrix with double-precision real values. You store the return values of the above function into **plhs[0]**, **plhs[1]**, etc. You can then use **mxGetPr** to obtain pointers to the actual data for your newly created variables, to store the return values there.

6. MATLAB does its own memory management. Avoid using C functions like **calloc**, **malloc**, **realloc**, and **free** for dynamic memory allocation; instead, use **mxCalloc**, **mxMalloc**, **mxRealloc**, and **mxFree** to avoid unexpected results.

7. Your gateway function must be named **mexFunction**, but any additional functions which do the "real" work may have any legal C function names, and there may be several such functions. For example, if your file **myfunc.c** contains C functions **mexFunction**, **func1**, and **func2**, at the MATLAB prompt, you can only call **myfunc** — you cannot call **func1**, etc. directly.

8. It is possible to enable your C code to use 64-bit indexing so that it can work with arrays too large for 32-bit indexing. To do so, use **mwSize** and **mwIndex**

(rather than **int**) as the data types of variables used to describe the size of an array or to index an array, and use the **-largeArrayDims** switch with the `mex` command when compiling.

14.2.2   Printing, warnings, and errors

Use the `mexPrintf()` function to print messages from inside your C code. This function behaves like **printf()**; you can either give it a simple string, or a formatting string followed by additional arguments.

To display a warning, you can call `mexWarnMsgIdAndTxt("ErrorID", "Some warning text")`. To print an error and exit your MEX code, call `mexErrMsgIdAndTxt("ErrorID", "Some error text")`. For both routines, the first string should be what is called a "message identifier," a tag identifying the source of error to MATLAB that can be used with various error reporting tools. It should consist of a "component:mnemonic" pair separated by a colon. More than one component can be used, separated by additional colons. For example, the message identifier "RMatlab:Cshuffle:nrhs" was used in **Cshuffle.c** to flag errors in the number of right-hand sides (input parameters). All substrings within the message identifier must begin with a letter. Subsequent characters can be letters, numbers, or underscores, and white space is not allowed.

The example below is the MATLAB equivalent of the R **testprint.c** function from Page 182.

```
────────────────── testprint.c ──────────────────
#include "mex.h"

void
mexFunction(int nlhs, mxArray *plhs[], int nrhs, const mxArray *prhs[])
{
  int n;

  n = (int)mxGetScalar(prhs[0]);
  if (n % 2)
    mexWarnMsgIdAndTxt("RMatlab:testprint:argAttributes",
                       "n was odd; it had value %d\n", n);
  if (n < 0)
    mexErrMsgIdAndTxt("RMatlab:testprint:argAttributes",
                      "n was negative!  It had value %d\n", n);
  mexPrintf("n=%d\n", n);
}
```

14.2.3   Random numbers

As in R, MATLAB maintains its own seed or state for its random number generators, which is independent of that used by the standard C library's **random** function. This is why the **Cshuffle** function was written to accept an optional third argument, which is used as a seed for the random number generator. It could also be modified to use the clock to see the random number generator if a special seed is provided, using code like that on Page 182.

There is no documented public API for calling MATLAB's internal private random number generators directly from within C; however, one can call any MATLAB function via the **mexCallMATLAB(nlhs, plhs, nrhs, prhs, funcName)** function. Arguments similar to those for the gateway routine must be used. For example, the following bit of code will generate an $n \times 1$ vector of random values from the exponential distribution with mean μ by calling MATLAB's **exprnd** function, print the generated values, and then compute and print their mean. A 1×1 matrix (i.e., a scalar) is set up with the desired mean, and then a 1×2 vector is initialized with the dimensions of the random vector to be generated. An **mxArray** is used to hold the results of calling **exprnd**, and another one for the results of calling **mean**. Note that memory is dynamically allocated to store the results of calling **mexCallMATLAB**; the memory is deallocated when the MEX file exits. You can use **mxDestroyArray** to manually free the memory earlier if desired.

```
─────────────── testmexcall.c ───────────────
#include "mex.h"

/* Two optional parameters can be given:
 *    n: how many exponentially-distributed random values to generate
 *    mu: their mean
 *    Default values: n=5, mu=2.0
 */
void
mexFunction(int nlhs, mxArray *plhs[], int nrhs, const mxArray *prhs[])
{
  double mu=2.0, *dblPtr;
  int n=5, i;
  mxArray *muArray, *sizeArray, *randValsMxArray, *rhs[2], *meanMxArray;

  if (nrhs >= 1) {
    n = (int)mxGetScalar(prhs[0]);
    if (nrhs == 2)
      mu = mxGetScalar(prhs[1]);
    if (nrhs > 2)
      mexErrMsgIdAndTxt("RMatlab:testmexcall:nrhs",
                        "One or two arguments needed: n mu");
  }

  muArray = mxCreateDoubleMatrix(1, 1, mxREAL);
  dblPtr = mxGetPr(muArray);
  dblPtr[0] = mu;
  sizeArray = mxCreateDoubleMatrix(1, 2, mxREAL);
  dblPtr = mxGetPr(sizeArray);
  dblPtr[0] = (double)n;  dblPtr[1] = 1.0;
  rhs[0] = muArray;  rhs[1] = sizeArray;
  mexCallMATLAB(1, &randValsMxArray, 2, rhs, "exprnd");
  dblPtr = mxGetPr(randValsMxArray);
  for (i=0; i < n; i++)
    mexPrintf("  value = %g\n", dblPtr[i]);
  mexCallMATLAB(1, &meanMxArray, 1, &randValsMxArray, "mean");
  dblPtr = mxGetPr(meanMxArray);
  mexPrintf("mean = %g\n", dblPtr[0]);
}
```

Using **mexCallMATLAB** to generate random numbers is highly discouraged because of the large performance cost. An alternative version of the **Cshuffle** function is given below, which uses MATLAB's function **randi** to generate the random values. The **mexFunction** gateway function is identical to the earlier version, and is not shown again here. This version runs approximately 25 times slower than the original version of the code on Page 188 which uses the random number generator from the C library. If you need to generate random values from different probability distributions in your C code and you are at all concerned with performance, you are much better off using standard techniques (for example, see References [5, 23]) to produce the values directly from within C than calling MATLAB code to produce them. If you know how many random values you will need ahead of time (or have a reasonable upper bound for the number), another alternative is to construct a vector of random values in MATLAB and pass that vector as a parameter to the C function. You can either produce a vector of values from a known probability distribution, or just produce uniform random values and then let the C code transform them into values from the desired distribution(s); for example, the **Cbirthdeath.c** function below transforms uniform random values (generated within the C code itself) into random values following the exponential distribution.

```
_____ Cshuffle2.c _____
void Cshuffle(double *outv, double *v, int n)
{
  int i, j;
  double tmpDbl, *imaxPtr, *retValPtr;
  mxArray *imaxMxArray, *retValMxArray;

  imaxMxArray = mxCreateDoubleMatrix(1,1, mxREAL);
  imaxPtr = mxGetPr(imaxMxArray);
  for (i=0; i < n; i++)
    outv[i] = v[i];
  for (i=n; i > 1; i--) {
    /* Generate a random integer from 0 to i-1
     * (not from 1 to i, because vectors in C use 0-based indexing). */
    *imaxPtr = (double)i;
    mexCallMATLAB(1, &retValMxArray, 1, &imaxMxArray, "randi");
    retValPtr = mxGetPr(retValMxArray);
    /* subtract 1 since randi gives vals from 1..i */
    j = (int)(*retValPtr) - 1;
    mxDestroyArray(retValMxArray);  /* free memory from mexCallMATLAB */
    /* swap outv[j] with the last element of outv */
    tmpDbl = outv[j];  outv[j] = outv[i-1];  outv[i-1] = tmpDbl;
  }
}
```

A C version of the birth-death simulation which uses C's **random()** function to generate all of its needed random values is below.

```
_____ Cbirthdeath.c _____
#include "mex.h"
#include <stdlib.h>  /* for declaration of random() */
#include <math.h>    /* for declaration of log() */

/*
 * phi, mu, s are input parameters.
 * eventTimes, popVec are used to return data
 */
void birthdeath(double phi, double mu, int s,
```

```
                        double *eventTimes, double *popVec)
{
  int i;
  double u, r, birthProb;

  birthProb = phi/(phi+mu);
  eventTimes[0] = 0.0;  popVec[0] = 100;
  for (i=0; i < s-1; i++) {
    /* generate a random value from exponential distribution
     * with mean 1/(popSize*(phi+mu)) */
    /* First generate uniform (0,1] value, being careful
     * not to generate the value 0 */
    u = ((double) random()+1.0) / (double)(RAND_MAX+1.0);
    r = -1.0/(popVec[i]*(phi+mu))*log(u);
    if (random()/(double)RAND_MAX < birthProb)
      popVec[i+1] = popVec[i] + 1;
    else
      popVec[i+1] = popVec[i] - 1;
    eventTimes[i+1] = eventTimes[i] + r;
  }
}


/* The gateway function.
 * Arguments are phi, mu, steps
 * Return values are eventTimes, popVec
 */
void mexFunction( int nlhs, mxArray *plhs[],
                  int nrhs, const mxArray *prhs[])
{
  int s;
  unsigned int rseed;
  double phi, mu, *eventTimes, *popVec;

  /* Check for proper number of arguments */
  if ((nrhs < 3) || (nrhs > 4))
    mexErrMsgIdAndTxt("RMatlab:birthdeath:nrhs",
      "Three input args are needed, and 4th optional: phi mu s [seed]");
  if (nlhs != 2)
    mexErrMsgIdAndTxt("RMatlab:birthdeath:nlhs",
                      "Two output values are needed");

  phi = mxGetScalar(prhs[0]);
  mu = mxGetScalar(prhs[1]);
  s = (int)mxGetScalar(prhs[2]);
  if (nrhs == 4) {
    rseed = (unsigned int)mxGetScalar(prhs[3]);
    srandom(rseed);
  }
  plhs[0] = mxCreateDoubleMatrix(1,s,mxREAL);
  plhs[1] = mxCreateDoubleMatrix(1,s,mxREAL);
```

```
  eventTimes = mxGetPr(plhs[0]);
  popVec = mxGetPr(plhs[1]);

  birthdeath(phi, mu, s, eventTimes, popVec);
}
```

Here is the equivalent MATLAB code:

```
─────────────────── birthdeath.m ───────────────────
function [eventTimes,popVec]=birthdeath(phi, mu, s)
birthProb = phi/(phi+mu);
eventTimes = zeros(1,s);  popVec = zeros(1,s);
eventTimes(1) = 0.0;  popVec(1) = 100;
for i = 1:(s-1)
  r = exprnd(1/(popVec(i)*(phi+mu)));
  eventTimes(i+1) = eventTimes(i) + r;
  if (rand < birthProb)
    popVec(i+1) = popVec(i) + 1;
  else
    popVec(i+1) = popVec(i) - 1;
  end
end
```

And finally, a C version which uses **mexCallMATLAB** to use MATLAB's random number generators to produce the needed values from the exponential and continuous uniform distributions. The **mexFunction** gateway function is identical with the one in **Cbirthdeath.c**, and is not shown again here.

```
─────────────────── Cbirthdeath2.c ───────────────────
void birthdeath(double phi, double mu, int s,
                double *eventTimes, double *popVec)
{
  int i;
  double u, r, birthProb, *expParmPtr;
  mxArray *expParmMxArray, *randValMxArray;

  birthProb = phi/(phi+mu);
  eventTimes[0] = 0.0;  popVec[0] = 100;
  expParmMxArray = mxCreateDoubleMatrix(1, 1, mxREAL);
  expParmPtr = mxGetPr(expParmMxArray);
  for (i=0; i < s-1; i++) {
    /* generate a random value from exponential distribution
     * with mean 1/(popSize*(phi+mu)) */
    expParmPtr[0] = 1.0/(popVec[i]*(phi+mu));
    mexCallMATLAB(1, &randValMxArray, 1, &expParmMxArray, "exprnd");
    r = mxGetScalar(randValMxArray);
    mxDestroyArray(randValMxArray);
    eventTimes[i+1] = eventTimes[i] + r;
    mexCallMATLAB(1, &randValMxArray, 0, NULL, "rand");
    r = mxGetScalar(randValMxArray);
    mxDestroyArray(randValMxArray);
    if (r < birthProb)
```

Method	Raw times		Times faster than MATLAB code	
	User	Elapsed	User	Elapsed
Cbirthdeath	0.14	0.141	600	594
Cbirthdeath2	104	104	0.805	0.805
birthdeath.m	84	84	1	1

TABLE 14.5

Timing results (in seconds) of simulating the birth-death process for $s = 3 \times 10^6$ events with $\phi = 1.1$ and $\mu = 1$ using three different methods. The C code (with native random number generation) runs 600 times as quickly as the MATLAB version.

```
      popVec[i+1] = popVec[i] + 1;
    else
      popVec[i+1] = popVec[i] - 1;
  }
}
```

Timings of the C and MATLAB versions of the birth-death simulation are shown in Table 14.5. The first C version of the simulation runs roughly 600 times faster than the interpreted MATLAB code. However, note that using **mexCallMATLAB** to generate the random values causes the C code to run even more slowly than the native MATLAB simulation, further evidence that making many calls to MATLAB from C code is a poor idea if performance is a main concern.

Finally, if you are considering using C to speed up your MATLAB code, you should consider obtaining MATLAB Coder from The MathWorks. This produces C or C++ code from MATLAB code, which lets you accelerate parts of your programs, or even build stand-alone executables from them.

Bibliography

[1] Yair M. Altman. *Accelerating MATLAB Performance: 1001 Tips to Speed Up MAT-LAB Programs*. Chapman & Hall/CRC, 2014.

[2] Stormy Attaway. *MATLAB: A Practical Introduction to Programming and Problem Solving*. Butterworth-Heinemann, 2013.

[3] Patrick Burns. *The R Inferno*, 2011. Available at http://www.burns-stat.com/documents/books/the-r-inferno/.

[4] Steven C. Chapra. *Applied Numerical Methods with MATLAB, 3rd edition*. McGraw-Hill, 2012.

[5] Luc Devroye. *Non-Uniform Random Variate Generation*. Springer-Verlag, 1986.

[6] Richard Durstenfeld. Algorithm 235: Random permutation. *Communications of the ACM*, 7, 1964.

[7] Laurene V. Fausett. *Applied Numerical Analysis Using MATLAB, 2nd edition*. Pearson Prentice Hall, 2008.

[8] R.A. Fisher and F. Yates. *Statistical Tables for Biological, Agricultural, and Medical Research, 3rd ed.* Oliver & Boyd, 1948.

[9] Amos Gilat. *MATLAB: An Introduction with Applications*. Wiley, 2014.

[10] Amos Gilat and Vish Subramaniam. *Numerical Methods for Engineers and Scientists, 3rd edition*. Wiley, 2013.

[11] Duane C. Hanselman and Bruce L. Littlefield. *Mastering MATLAB*. Prentice Hall, 2011.

[12] Desmond J. Higham and Nicholas J. Higham. *MATLAB Guide*. SIAM: Society for Industrial and Applied Mathematics, 2005.

[13] The MathWorks Inc. *MATLAB External Interfaces*, 2012.

[14] Owen Jones, Robert Maillardet, and Andrew Robinson. *Introduction to Scientific Programming and Simulation using R*. Chapman & Hall/CRC, 2009.

[15] Donald E. Knuth. *The Art of Computer Programming vol. 2, 3rd ed.* Addison-Wesley, 1998.

[16] Friedrich Leisch. *Creating R Packages: A Tutorial*, 2009. Available at http://cran.r-project.org/other-docs.html.

[17] G.R. Lindfield and J.E.T. Penny. *Numerical Methods using MATLAB, 3rd edition*. Academic Press, 2012.

[18] Peter Linz and Richard L.C. Wang. *Exploring Numerical Methods: An Introduction to Scientific Computing using MATLAB*. Jones & Bartlett, 2003.

[19] John H. Mathews and Kurtis D. Fink. *Numerical Methods Using MATLAB, 4th edition*. Pearson Prentice Hall, 2004.

[20] M. Matsumoto and T. Nishimura. Mersenne twister: A 623-dimensionally equidistributed uniform pseudo-random number generator. *ACM Transactions on Modeling and Computer Simulation*, 8:3–30, 1998.

[21] Paul Murrell. *R Graphics, Second Edition*. CRC Press, 2011.

[22] William Palm III. *A Concise Introduction to MATLAB*. McGraw-Hill, 2007.

[23] William H. Press, Saul A. Teukolsky, William T. Vetterling, and Brian P. Plannery. *Numerical Recipes in C: The Art of Scientific Computing*. Cambridge University Press, 1992.

[24] R Core Team. *Writing R Extensions*. R Foundation for Statistical Computing, Vienna, Austria, 2012. Available at `http://cran.r-project.org/manuals.html`.

[25] R Core Team. *R Data Import/Export*, 2013. Available at `http://cran.r-project.org/manuals.html`.

[26] R Core Team. *R Language Definition*, 2013. Available at `http://cran.r-project.org/manuals.html`.

[27] Timothy Sauer. *Numerical Analysis, 2nd edition*. Pearson, 2012.

[28] W.N. Venables, D.M. Smith, and R Core Team. *An Introduction to R*, 2012. Available at `http://cran.r-project.org/manuals.html`.

Index of R commands, variables, and symbols

*, 43
->, 10
..., 88
.C, 178, 180–182, 185, 188
.Call, 180, 181, 185, 186, 188
.External, 185–188
.First, 172
.Last, 173
.Last.value, 13, 84
.Machine, 138
.RData, 161, 172
.Rprofile, 172
.packages, 173
/, 44
:, 20
;, 11
<, 24
<-, 10, 85
<<-, 37, 87, 88
<=, 24
=, 10, 85
==, 24, 168
>, 23, 24
>=, 24
?, 5
??, 5
[, 22, 27, 57
[[, 57
#, 80
$, 57
%*%, 43, 44
%%, 13, 14
%^%, 44
%in%, 171
&, 49
&&, 65
^, 13
{, 66, 68
|, 49
||, 65
!, 49
$, 36
^, 44

abs, 13, 16
acos, 13
acosh, 13
alarm, 170
all, 49, 66
any, 49
apply, 41–43, 47, 50–52
approx, 140
apropos, 5
Arg, 16
array, 32
arrayInd, 29
arrows, 116
as.character, 167
as.Date, 170
as.formula, 141
as.integer, 157
as.list, 60
as.matrix, 148, 149, 151, 156
as.POSIXct, 170
asin, 13
asinh, 13
assign, 16
atan, 13
atan2, 13
atanh, 13
attach, 36
attributes, 167
axis, 120

beta, 15
bquote, 115
break, 72, 73
browser, 172

c, 20, 22, 28, 34
cat, 166
cbind, 20, 22, 26
ceiling, 13, 14
chol, 45
choose, 15
class, 17, 57
close, 148

201

Index of MATLAB commands, variables, and symbols